The Invention of Crowdfunding

A Circuitous Tale of How Marillion, Their Crazy Fans and I Accidentally Upended the Music Industry

by
Jeffrey Pelletier

Published in the United States by IngramSpark, La Vergne, Tennessee.

Printed in the United States of America.

ISBN 979-8-218-16101-9

Photo Credits: Jeffrey Pelletier, Julie Hanlon, Andre Pelletier, Ellie Chappell, Mark Hanlon, Mike Hanlon, Molly Burr, Maggie Surette, Andy Saks, Stephen Humphries, Kim Proctor, Kim Robinson, Erik Foxvog, Brian Steed and Lily Zampitella.

All copyrighted Marillion artwork and lyrics used by kind permission of the band and manager Lucy Jordache.

Marillion cover artwork and photos by Mark Wilkinson, Bill Smith, Carl Glover, Sarah Ball, Paul Cox, Andrew Gent, Niels van Iperen, Matt Curtis, Darren Newitt and Simon Ward.

Find a better way of life at www.marillion.com

Edited by Mike Hanlon.
Chapter heading crowd silhouette designed by Kjpargeter / Freepik.
Chapter banners designed by Mike Hanlon.
Story banners designed by Mike Hanlon and Jeffrey Pelletier.

First Edition

Tables of Contents

The Invention of Crowdfunding

Continued on next page

Tables of Contents (Continued)

And Other Stories

Continued on next page

And Other Stories (Continued)

Recipes

The Invention of Crowdfunding

Prologue

This isn't a dream. Neither is it a daydream, nor a simple flight of fancy. This is actually happening.

I'm standing at the front of the Carolina Theater in Durham, NC in 2018. The lead singer of my favorite band, Marillion, has stopped the show momentarily and is holding me tight with a heartfelt embrace. Behind me, a thousand fans are standing and cheering... for me. Even the rest of the band (I was later told), are applauding and smiling, *genuinely*.

How did I get here? What improbable set of events could possibly have led to this moment? How can it be that all these people know who I am and remember the small role I played in the history of this band we all love? I am quite literally overcome with emotion and gratitude, but we'll get to that in much more detail later.

When I eventually came down from that surreal high, I realized something. I had to tell this story. Whether I told it to an audience of ten or ten thousand, I needed to get this out there. I knew that I would write a book about what brought me to the front of that stage, but it took me a year before I would begin the actual writing process. All the while that writer's itch inside me kept demanding to be scratched. Eventually, in 2019, I couldn't *not* try to tell this story any longer. I opened Microsoft Word and clicked on "Blank Document."

I didn't want to just tell a linear tale of how Marillion came to be known as the recipients of the first successful online crowdfunding campaign. I wanted to tell the story of what led me to the band. Of what music has meant to me. Of what the people in my *life* mean to me. And hopefully, arrive at that moment with the reader truly knowing me – and understanding how much it meant to me.

So, if that's a journey that you might want to start with me, then take this book, and let's go together. Much like the varying passages that synergistically make up an epic Marillion tune like "Ocean Cloud" or "This Strange Engine," this will be a twisting tale that will take a few detours before we get to our final destination. I hope you enjoy the ride.

Introduction

Hi there. I'm Jeff.

Thanks for buying this book. Or conversely, thanks for supporting your local library. Or, short of that, umm… thanks for saving a milli-tree by borrowing this from a friend and requiring one less book-sized allotment of wood pulp for these pages.

OK. Who are we kidding? You're probably reading this on a screen of some sort. So, we both know that that last one was a stretch. But still… *Thank you.* If you're actually going to spend the time to listen to me prattle on for several hours, then I appreciate that. Truly. Even if I don't make a cent or a shilling from it. I hope I earn the honor of your time and your attention during this time we'll spend together.

So again, I'm Jeff. Hi. Nice to meet you. What? That's not what a book introduction is? I think maybe that word doesn't mean what you think it means then. It's an introduction. So, you *introduce* yourself, right? Wrong? Well, tough.

Right then. Hi. Jeff here still. Jeffrey L. Pelletier to be exact. This is me.

Well, that's me if I was a character on The Simpsons. Here's me at 5? And again turning 50.

The L stands for Louis, which mainly comes from my dad's late brother (my godfather)—and I'm guessing also from a rich French historic tradition. (*Abbey Road* plays in the background... *He-ere... Comes... the Su-uuun... King*). Remember all those Louis's? I'm guessing my parents did too. Speaking of French history, did you know that the infamous highwayman Jacques Pelletier was the first person ever killed by guillotine? C'est vrai. We Pelletiers have a lot to be proud of.

Hello! Did I introduce myself yet? I'm Jeff. And what exactly is this disorganized mess that you're holding in your hand? Let's face it. It took me over 50 years to write one book. So, chances are this is the only one I'm ever going to write. Because of that, I'm going to pack everything I can into this sucker. I know that's not how it's typically done. You normally have to stick to one genre. Otherwise, what section of the bookstore are they going to put the thing in? (*What's a bookstore*, you ask? Oh, man. What's happened to our world?)

In writing this, I didn't want it to read like a normal book. I wanted to avoid the monotony of a stream of *this happened*, then *this happened*, then *this happened*, etc.... My goal was to make this feel like a conversation between me and you (the reader). Accordingly, I'm going to jump back and forth between our main narrative (the history of crowdfunding and the band Marillion) and some personal stories. I'll be going off on tangents here and there. You can join me on these diversions. Or you can skip ahead whenever you're so motivated.

And that's perfectly OK. You see, you're an autonomous being. You have free will. If you decide you're not interested in crowdfunding or anything to do with that British prog band I'll keep talking about over and over again, you can just read the childhood stories. Not interested in those old stories? Just stick to the crowdfunding narrative (by reading only the "Chapters"). Or just skip to the extra stuff at the end. I broke them all out separately in the Tables of Contents, in case you breezed past that and didn't notice.

So anyway... We have a lot of history to get through if you're interested in internet-based crowdfunding and how it first came to be. And interspersed within that chronicle are a lot of unrelated stories that I hope you'll enjoy. And what the heck. I'm going to throw in some of the things I've picked up along the way – things I've learned by not dying for over five decades on this planet and experiencing life among the humans.

So read the whole thing. Or just skip to the parts that you fancy (or *think* that you'll fancy). I won't be insulted. It's your book now. You can use it as you like. But again. I'm Jeff. Hi. And thanks again for reading this book. Or... thanks for at least *starting* to read this book. I'm not going to immediately saddle you with that kind of pressure.

OK. Now that the pleasantries have been handled, let's get to it...

Chapter 1

Invention?... Seriously?

Let me start by saying what I often say when asked about inventing crowdfunding. *Yes*. It's pretty silly to think that someone could *invent* the concept of many people pitching in for something collectively. You won't get any arguments from me about that. What I can tell you is this…

I was lucky enough to be the guy who said the right thing at the right time in the right place. And after that, a little bit of history took place.

Chapter 2

What the Hell is a List Server?

In January of 1997, Mark Kelly (the keyboardist for the UK band Marillion) announced on an e-mail list server dedicated to the band that they once again would not be able to tour the U.S. due to losses incurred on previous tours.

Wait. Hold on. What the hell is a *list server*, you might ask? Well, if you don't remember a time before Coldplay, allow me to explain.

Back in the olden days of the late 20th century, the internet was still a novelty. Those of us at work in the 90's, (who had these new communication mechanisms called e-mail addresses), often didn't even have access to the actual internet yet. And then when we *did* get access, we initially had to submit for approval for each website we needed to visit. So obviously, chat rooms weren't a possibility at work. Yet if you couldn't join an internet chat room, one thing you *could* do was subscribe to a list server.

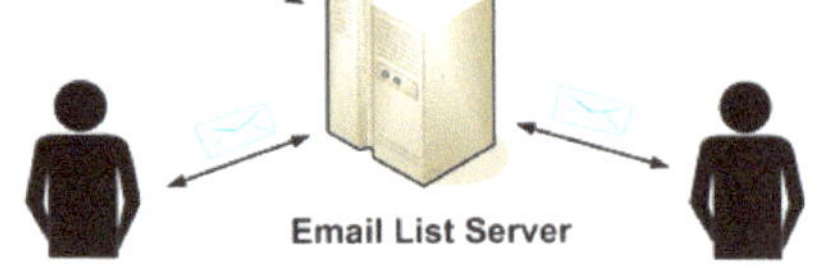

A list server was an e-mail service you could join. Each list server was dedicated to a specific topic. Once you joined a list, you would get a giant e-mail each day called a "Digest" from that group. This digest compiled all the e-mails each subscriber sent in to the group that day into one big e-mail (so your mailbox wasn't inundated with all the separate messages). You could just read them all as a "lurker" or you could join in the discussion and pick and choose which topics to reply to. If the group was extra busy, you would get multiple digest e-mails that day.

This technology did have its pitfalls. Sometimes people would reply to the digest and include a complete copy of the previous digest with their reply. Imagine having to scroll through the equivalent of a dozen pages of text just to see if there was room for another message at the end. And topics weren't collated at all. It was just a long list of emails displayed chronologically. So, as you scrolled through this long missive, the subjects would change chaotically. Some people quoted what they were replying to *above* their response; others replied and had the original text below. It took a bit of linguistic gymnastics to navigate the discussions.

This was still an extremely exciting thing at the time though. You see, up until this point, Marillion had been a completely *private* passion for me. A lot of my friends *knew about them* because I never stopped talking about music, and Marillion was my favorite lesser-known band. But it always seemed like I was the only one in the world who loved them this much.

Then suddenly this new forum allowed people with the most eclectic passions to find each other and talk about these private loves that they shared. It was a revelation, to suddenly have a whole new world opened up to us and not to feel alone anymore. We talked about this great band that we loved. We argued about which songs were best and which lead singer we preferred. We dissected the meanings in the lyrics and talked about our never-ending (and seldom-successful) attempts to convert new fans to this amazing music.

So back to the Marillion list server... It was called the *Freaks* list. This is an allusion to the same-named Marillion song in which they sing, "*All the best freaks are here, please stop staring at me...*" We had somewhere in the range of 1,500 to 2,000 fans around the time I joined. And how cool is this? A couple of the band members actually posted to this list occasionally. *Holy crap. Is that actually Mark Kelly who wrote that? And he's taking the time to join the discussion. Wow!*

What a time this is, everything changing
Faster than the eye can blink, faster than we can stop and think
What will the future hold?

- Lyrics by Steve Hogarth

Chapter 3

Blame the Eagles

So, as I was saying, in 1997 Mark Kelly said... No, wait. We should really establish something first.

In May of 1984, just before I turned 17, I saw my first rock concert. It was the *90125* tour of the band Yes. As of this writing 37 years later, I have seen 396 concerts (not including theater shows, stand-up shows, or local club bands). So, you can pretty easily conclude that I was hooked on live music.

We made the trek all the way out to Worcester, Mass., from our little eastern Massachusetts town of Andover. That night my adolescent brain was just blown away. I had never heard music played so *loudly* and had never been amongst so many like-minded people responding so *enthusiastically* to this thing that previously had only existed in my room and blasting from the car stereos as we would drive around town.

And it didn't hurt that Yes is one of those bands that sounds *fantastic* live. I could feel my chest being physically pushed into my body from the pounding bass notes in "City of Love." Jon Anderson's otherworldly, high-pitched voice resonated strongly throughout the arena—reaching an almost transcendent level on "Soon," the opening section of "The Gates of Delirium." The lights were choreographed to the changes in the music, adding a wonderfully unexpected emphasis to my emotional response to the sound.

During the song "Hearts" they even had translucent screens suspended over the crowd. Their laser lights drew giant hearts on them, making it seem like these hanging holograms were hearts of hovering light floating above us. This was something I knew I wanted to be a part of again and again and again.

Over the next decade, I saw 81 more concerts. That first Yes concert cost me a whopping eleven bucks for a ticket ($11.50 to be exact). And I remember being kind of pissed that I had to pay a one-dollar fee to *Ticketron*. (Little did he know... Ha. Hey, *TicketMaster*. You *suck*.)

I regularly paid between 10 and 30 dollars per ticket over those ensuing years. And I believe the most expensive show I had seen to that point was Pink Floyd at the old Foxboro stadium, which cost about 35 bucks.

And then along come the frigging Eagles.

You see, up until that point, no one ever thought that you could charge 100 dollars a ticket and get away with it. It was pretty much understood that people wouldn't pay that much for a concert ticket en masse.

An Eagles reunion had been the "white whale" of desired rock shows. The clash of personalities amongst the band members was well documented. They had broken up and they had said it was for good. Fans had been clamoring for a reunion for years, yet the band would never consider it... *Until they finally did*. So, when they in fact did reunite in the 1990s for an actual Eagles tour, the aptly-named "Hell Freezes Over" tour was born.

It was the first tour I can remember where the price of the ticket correlated to how good your seats were. And not only was it the first tour to charge that unheard of $100 per ticket, but the top tier tickets were $120. (The tiers were $80, $100 and $120, if I remember correctly.)

So, if you look at the cost of concert tickets today (that regularly cost *hundreds* of dollars), and you're looking for someone to blame... *Blame the Eagles*. (Hey, Don and Glenn. Thanks a frigging *lot*. You guys are great, but it's all your fault. Seriously. You screwed us all.)

$80 for the cheap seats. Unheard of at the time.

Note that we lucked out when an usher came through the last two rows at Great Woods and relocated all of us about 15 rows closer, to the back of the $100 level that didn't sell out. Melissa and I were moved from Row CC (the last row in the pavilion) up to Row N of the back sections. We were still over 40 rows from the stage, but what a difference that made compared to being in the very last row. (Melissa Z has been a fixture in my life since adolescence and will get a proper introduction in the pages to come.)

Chapter 4

That First Taste

So again… in 1997, Mark Kelly announced on the Freaks List - No, wait again. Hold on. Let's go back one more time before picking up there.

In 1995 I saw my very first Marillion concert at the Paradise Rock Club in Boston. It turns out, I could have seen them in my city before. But in the days before the internet, I just had no *idea* they were here. What a shame. If the Boston Globe arts section or if the local rock stations didn't mention a show, it just passed us by. And we were never the wiser. Well, not 'til a few short years later when you could look up any piece of information in recorded human history (including every date your favorite band ever played, or *would play* in the near future).

The scary thing was how close I came to missing this Marillion show as well.

Each Thursday in the aforementioned Globe, there was an insert called the *Calendar* section. It focused mainly on the arts. I remember coming home from work on a Friday and starting to throw Thursday's paper in the recycle bin. I noticed that I hadn't read the *Calendar* section yet and put that aside.

Julie was currently in her car making her monthly trek back from Connecticut for the weekend. This was during the years when she was working at Yale, before she moved back home in '97 to work at Harvard. I put on some music, as always, and sat down with Friday's paper and Thursday's *Calendar*. I opened *Calendar* and was thrilled to see a review of the new Marillion album, *Afraid of Sunlight*.

How cool is that! There was even a picture of the album cover artwork at the top of the review with the winged angel boy standing before that ring of fire.

I read the review in a hyper-excited state. I think I remember it being a generally positive review and wondered if this might inspire a few people to discover this great band. I got to the end of the review and there was a line that stopped me in my tracks. "*Marillion play The Paradise tomorrow night.*"

What? Read it again. *Tomorrow night.* Tomorrow night… spoken *from yesterday.* That's *tonight. Holy Crap!!!* I frantically called the number for The Paradise listed in the paper and let it ring about 18 times. (*Again, the days before the internet*). Someone actually answered the phone eventually. I breathlessly asked if Marillion were really playing and if there were still tickets available.

The person on the other end just laughed and said, "*Oh yeah.*" The club held fewer than a thousand people, but it was not going to be a sellout. Oh my God. Oh my God. *Oh my God!*

Julie's coming tonight. She isn't going to want to go. I can't miss this. She's gonna be so pissed at me. Crap, crap, crap, crap, craaaaaaaap!

Her trusty Honda Civic eventually pulled into my driveway. Out popped Julie, tired from a two-and-a-half-hour drive, but happy to see me. I open the door and words flew out of my mouth so fast she probably didn't hear half of them… "Marillion is playing tonight! There's still tickets. We have to go! Well, *I* have to go, but you don't have to! But it's *Marillion*! Can you believe it?!!"

So yeah. She understood enough of that to get the gist of it. And her face just dropped into a dejected expression most accurately articulated as… *Ugh.*

Story 1

At this point, let's take a little break from our history lesson.

I've always been a fan of nostalgic storytelling–tales of innocence, or the loss thereof, tales of simpler times… Think back to that fantastic pilot episode of (the original) *The Wonder Years*, with Kevin Arnold navigating the shifting world of the late 1960's, neighbor kids going off to war, friendships growing and changing and relationships with the opposite sex redefined–an episode which culminated for Kevin in a heartwarming kiss from Winnie Cooper. For me, storytelling doesn't get much better than that.

A number of years ago I started to document some of those incidents that resonated with me along that same sentimental wavelength. I never thought I'd collect enough tales to fill a book. And I never really thought I'd have enough interesting things to say to make that non-book worth publishing. But here we are. And one thing I know for sure is that there won't be a repeat occurrence of this.

So, what you have here is a collection of everything worth talking about, 'cause as I said, this is my only chance to put these stories out into the world. And what could be more in tune with that desired nostalgic tone than a child's love of candy? Thus, I present you with the first of our little journeys together in my personal Way-Back Machine. Here is *Candy Day*.

Age 5, Grade 1, 1973
"Candy Day"

Every Friday was "Candy Day." Candy Day was the culmination of the week and something that made the whole day as exciting as summer vacation. Each Friday morning, Mom would give my older brother Alan and me a nickel each to buy a candy bar on the way home from school. This was back before Hershey decided that their chocolate bars were worth a dime and all the other candy bars shortly followed suit.

Alan and I walked almost a full mile to our Catholic grammar school—first down Shawsheen Road, and then up the big hill towards St. Augustine's elementary. We passed the little store next to the railroad tracks on the way to school, but we were good and we always waited until after school to go in and use our newly acquired wealth.

All day long that nickel felt like it was burning a hole in my pocket. I couldn't wait to get out of school and it was taking *so long*. Finally, the day was over. The last bell sounded and my brother and I raced each other to the store.

I kept my books in the same canvas bag that just about all the kids in our school used - the dark green one with the long strap. Those bags could take a beating. Mine bounced and skidded along the sidewalk as I dragged it behind me with outstretched arm.

We made it to the store and scrambled inside, shattering the store's peaceful afternoon tranquility. The old man behind the register had probably gotten to know our schedule and I thought he might even have been awaiting our arrival.

"Easy now, boys," he said. "The candy's not going anywhere."

Alan immediately picked his favorite—a Milky Way bar. My favorite was Snickers. I liked Milky Way too. But since we hated each other, I wouldn't get the same candy bar as he did if I could help it.

I was just about to grab the old standby Snickers when a bright orange wrapper caught my eye. It was reflecting the light just right and seemed Oh-So-Yummy, just waiting for me. I grabbed the Reese's Cups, thinking about how I would eat them. First, I'd nibble around the outside, eating the entire ridged perimeter, leaving the round chocolate/peanut butter/chocolate sandwich, which I would then pop into my mouth whole.

My brother paid for his first. Then I paid for my orange prize and stepped back outside. Alan had already decided to head home at top speed without waiting for me (something he *wasn't* supposed to do).

Jerk, I thought, and bolted after him.

As I headed down the big hill, my body started to go a little faster than my tiny legs could keep up with. My book bag bounced and skipped behind me then hit the curb and (to my shocked disbelief) wrapped itself around my legs. I got to take two more constricted mini-steps before I flew forward, completely horizontally.

I reached out with my right hand to break the fall, but the strap of my book bag held it back. I reached out with my left hand but at the last second noticed that this was the hand that held my Reese's Cups.

NO!, I screamed inside my head and landed headfirst on the sidewalk. My face bounced and skidded across the dirty pavement grinding sand into my cheeks.

The momentum of my landing brought my left arm down hard against the ground. In a last-ditch effort to save my candy bar from being crushed, I bent my hand up at the wrist. My arm slapped the sidewalk at the wrist and my fingers reflexively flew open, sending the package of candy cups forward through the air.

At this point, Hollywood special effects take over my memory. In slow motion, I saw the bright orange-wrapped candy bar rotate once in the air, bounce on its corners twice and skid (right side up) across the sandy sidewalk, over the curb and then straight down a sewer drain.

My body dragged itself to a stop next to that very sewer grate. I lay there for a minute staring at my rusty nemesis, not even believing what had just happened. Then the inevitable juvenile reaction came... I bawled my little eyes out.

These were the days after the divorce (but before my mother had to make that impossible choice between staying with her kids or leaving with the man she loved). So, I thought she was too poor to give me another nickel, raising two kids on her own. I even imagined she might punish me for wasting the one she had already given me. And dammit, my face was all scraped up and it *really* hurt.

Walking the rest of the way home alone, pouting like a real pro, I made a vow to keep this to myself. I didn't tell anyone about the whole incident until years later. But don't cry for me, Argentina. Just do me a favor. If you ever get your hands on a time machine, go back to that fateful Andover spring day in 1973 and give that little kid another nickel. Or even better, catch me before I fall.

I haven't eaten a Reese's Cup since without remembering that damn thing turning through the air in slo-mo.

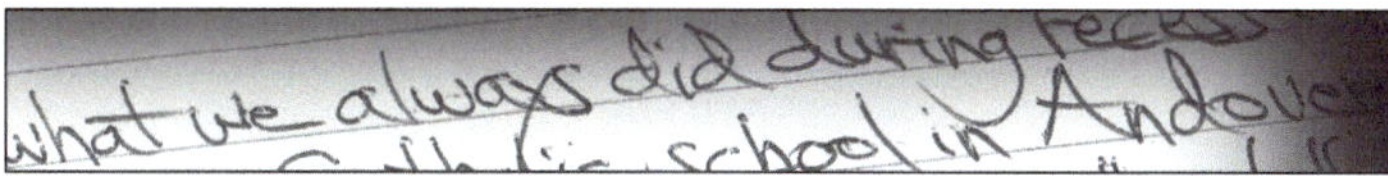

Story 2

In this, our second installment of musings on more innocent times, we explore a three-year-old's ability to make friends in an instant, along with an inspiring display of athleticism.

Look. I got it in the air!

No. We're not going to talk about kites. How many of you remember *Big Wheels*? Back in my childhood, this was probably the most popular toy for kids that were too young to ride bikes. Imagine a giant molded piece of plastic resembling a tiny car with no doors or roof. Instead of four wheels though, a Big Wheel had only three. One giant (i.e. "big") wheel in front was connected to your pedals (hence the name), and two little wheels in the back were just along for the ride.

If you tried to pedal too quickly, that big front wheel would spin uselessly before gaining traction. And if you turned the handlebars quickly, that wheel would slam perpendicularly to its limit and spin you out. Although not good for turning and maintaining speed, this was great for skidding to an impressive stop. Over and over, we would try to look cool by gaining what we thought was a remarkable speed in our *Big Wheels* and then jackknifing to a stop with a complete, crowd-pleasing rotation. We were kings of the dead-end road in our plastic suicide machines.

Here is how that *Big Wheel* led to me meeting my own Winnie Cooper.

Age 3, 1970
"The Big Wheel"

It was a beautiful summer day and Mom was sitting on the porch sipping a cup of hot tea. I couldn't understand why she would want to drink anything hot on a day like this, but hey... who could figure adults.

Across the street, an anxious little girl squirmed and fidgeted as the moving vans were unloaded. Finally, the thing she had been waiting for was brought out. She grabbed her Big Wheel, immediately hopped into the pilot's seat and pedaled her way to the top of the street.

Brechin Terrace was a short dead-end road in Andover, halfway down the hill from the High School. There were thirteen duplexes housing twenty-six families and spawning forty-three children (including Jessica, our newest addition).

Jessica sped down the street at an impressive pace, returned to the top and repeated her circuit. Three times she raced up and down, her blond curls streaming behind her. Then Mom innocently suggested, "Why don't you go introduce yourself to the new girl, Jeffrey."

I gullibly accepted her proposal. I walked out from my porch into the street on an intercept course with the three-wheeled demoness of my childhood. I reached the center of the road, turned ninety degrees to my left and held my arm out (palm up) with all the authority of a veteran traffic cop.

"Stop!" I commanded.

Jessica, in the meantime, must have seen me coming but had not changed her trajectory in the slightest. At the last moment, she kicked back on the pedals and spun her plastic vehicle sideways, coming to an impressive skidding stop with inches to spare.

My green eyes met her blue, and in a jolt of premonition I saw the next fifteen years course through me like electric fire. Sneaking out late at night to jump in the neighbors' pool; the time she kicked Andy Thompson in the shins after he knocked me flat with an unexpected right hook; our mutual first kiss in her backyard before either of us knew that there was more you could do besides kissing.

"Hi, I'm Jeffrey," I proudly proclaimed.

"I'm Jessica," she replied. "Got a Big Wheel?"

"Of course," I said.

"Well, go get it. Let's race."

"OK!"

I ran to the garage and brought back my own set of wheels. At the top of the street, we waited together while she recited the litany.

"On your mark... Get set... *GO!!!!!!*"

Our big front wheels spun around, scratching for a grip on the road. The sound of plastic friction against pavement resonated throughout my world. She pulled out to an early lead and I rode her tail, desperately trying to impress my new friend with a demonstration of my physical and athletic prowess. And (despite my desire to alter the facts...) I never even came close. She blew me away like I was standing still. I ate her dust. I never saw anything but her long blond hair waving me goodbye, as the distance between us grew larger and larger.

Looking back... isn't that whole incident just the *perfect* metaphor for every relationship I've ever had with a woman?

[*Author's Note: Well, until Julie came along. But that's a whole 'nother story.*]

Chapter 5

"He Sang the *Shit* Out of That Song!"

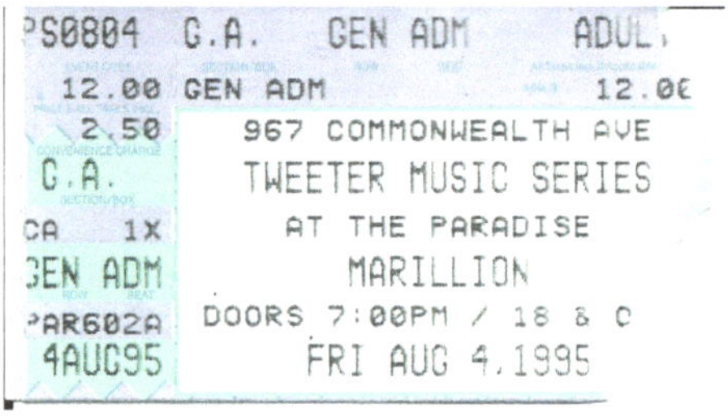

OK. Back to our Marillion tale. Where did we leave off? Mark Kelly and the *Freaks* List?... No. *Ugh*. Yes. That's where we were. "*Ugh*."

You see, I had tried unsuccessfully over and over again to indoctrinate Julie (and the rest of my friends) into this "cult of one" that I belonged to called *Being a Marillion Obsessive*. I'll give her credit. She listened with an open mind at least.

She had great taste in music, by the way. She was into The Police (and of course, Sting) before I even knew who they were. When she was only fifteen, she lied to her Mom—saying she was going to see a movie at the Lawrence Showcase and would head over to the Denny's across the street afterward. Instead, she then proceeded to hop into Teddy Silverman's rusty yellow Chevy Nova and was driven down to Providence, Rhode Island on a school night to see The Police on the *Ghost in the Machine* tour in 1982. (Pretty impressive act of rebellion there, Miss Goody-Two-Shoes.) And she'll never let our Rush-obsessed college friend, Dan Peduzzi, forget that she owned *Permanent Waves* before he had even bought his first Rush album.

Because of that good musical taste, I always thought I had a chance to convert her. Yet, while she admitted she was impressed with some of Fish's lyrics (with their evocative imagery), the songs never really found a latch in her brain to hook onto.

Afraid of Sunlight had been playing pretty regularly in my car, my home and my Discman (remember those?) since I got it a few weeks beforehand. And I was hoping I could get Julie into a positive enough mindset to enjoy this show. On the drive into the city, I put the CD into the player. I hit skip forward twice. And I asked her to *just give this song a chance, 'cause I really think you're going to like it.*

Track 3 started to play and I turned it up to that not-so-fine line between *not loud enough to fully appreciate it* and *blowing the windows out*. The name of the song was "Beautiful"—a gorgeous track that wasn't quite ballad, but wasn't quite rock song. It had a positive message (about appreciating the beauty inside all of us), and I had a feeling it was just the type of song that could finally win her over.

We crested the rise on Route 2 East where if you wanted to, you could shift your car into Neutral and coast down a steady slope for almost a full mile maintaining highway speed as the Boston skyline is laid out before you... And as we cruised down the long decline with the city lights in full view, Steve Hogarth sang out to us in an impassioned plea... *Why don't you stand up and say, Give yourself a break, They'll laugh at you anyway, So why don't you stand up and be... Beautiful...*

OK, she admitted as the song faded out. *That was pretty good.* I couldn't help but smile. This would indeed be a great night.

Marillion's name on the marquee (haphazardly displayed, or not) at the Paradise Rock Club in Boston.

We worked our way through the twists and turns of Memorial Drive along the Charles River and found our way to the club in the Brookline section of Boston where my musical heroes were actually going to be appearing. I walked in wondering how many people would be there and was both disappointed and pleased to see the floor was packed. Both of us being shorties, Julie and I worked our way up to the second level where we hoped to get an actual view of the stage. Up here people were lining the banisters. *Wow*. Does everyone here actually *know* Marillion? *Inconceivable...*

We found a spot to sit on top of a booth table where we could see the stage between a couple of the banister leaners. I was fidgeting with anticipation as we waited for the band to take the stage. And in that time, I kept thinking, *I can't believe I'm actually going to see Marillion*. Up until this point, they'd only existed inside whatever speakers I was hearing at the time. But now they were going to be just several yards away from us. *Will they sound as good as the albums? Will Steve Hogarth be able to do justice to the Fish songs? Will they even play any of the old songs? I've obsessively loved this band for almost ten years now and I'm finally about to see them live. Can they possibly live up to the expectations?*

The band takes the stage to an absolutely rocking version of the Fish-era classic "Incommunicado." H is belting out the intricate lyrics with a powerful voice and serious stage presence. He stands on top of the speakers at the front of the stage and has the crowd in his metaphorical hand. And the band sounds undeniably *incredible*. Any doubts I had about h's acceptance as the *new* singer, or of the band not living up to my expectations, were dispelled in that first song. "Incommunicado" ends and the crowd erupts in a joyous roar. All those years of this being a private passion are ended. I am amongst friends.

The people here, Full of love and comfortable in themselves
Not scared to let go, No fear round here.

- Lyrics by Steve Hogarth

Not long after that, they played an absolutely stunning version of the opening track from the new album, "Gazpacho." As h roared out the line *You can tell it to the ocean...*, I remember as he drew out that last word *oce-aaaaaaaaaan*, the entire building resonating with a thunderous voice that was so powerful and yet so beautiful. My jaw might have actually dropped open at the sheer power of musicality on display from both band and singer.

If I remember correctly this was followed immediately by our highway song "Beautiful." It sounded gorgeous and I was enthralled. When the song ended (just as I was thinking the moment couldn't be more perfect), Julie leaned over to me and yelled over the crowd into my ear, "He sang the *SHIT* out of that song!" I turned to see a beaming grin stretched across that lovely face.

Mission accomplished. Julie is now one of us. Resistance is futile.

To recap: Julie is blown away. I'm in ecstasy. The next two hours are a blissful reverie. Serious chunks of *Brave* and *Afraid of Sunlight* are played to their utmost brilliance. We even get rocking renditions of "Heart of Lothian" and "Garden Party" from the early days. And as Steve Rothery pulled out his acoustic guitar, I knew we were about to be treated to the song "Easter" – a song with possibly one of the greatest guitar solos ever recorded. This is as good as it gets (or so I thought at the time).

I have a very good friend from my college dorm days named Ed Murphy. He's just about the nicest guy you've ever met. And you wouldn't expect it from his buttoned-up wholesome appearance, but he's a total "prog head." He introduced me to Marillion, in fact (as well as King Crimson and several other progressive bands). Well, Ed had previously told me that when you see bands in clubs like these, sometimes they'll come out after the show and sign CDs. I had never done this before. But this show was different. Why not give it a shot?

I asked Julie if she was willing to stay and see if the band would come out and talk with the fans. Trouper that she is, she immediately agreed, even though we were going to a wedding the next day. Greetings to Lorraine Coughlin (now Lorraine Sabella), if you're reading this. During your wedding, all I kept doing was replaying the concert the night before over and over in my mind. But don't worry. The food was great. *And quality cake*. For the guests, all we care about is the cake.

And so, we agreed to stick around. And not long after, sure enough, out came h and Steve and Pete. Quick hellos and autographs were obtained from Steve R and Pete T. And we then decided to wait for the chance to meet the voice of Marillion. He signed my *Afraid of Sunlight* CD sleeve and gracefully accepted my compliments on this latest masterpiece.

Digging out this signed AOS sleeve today, I'm reminded of how afterward I stared at what h had written across the shiny suit on the right-hand page. For days I wondered, "why did he write EWIS?" I very stupidly didn't realize that of course, he had written "Elvis xx"—in honor of what I later learned was Elvis Presley's silver phoenix suit.

Julie then told h that he had made a new fan that night and how she resisted Marillion for so long, but now she couldn't wait to see them again. She leaned over and kissed him on the cheek. Julie then pointed to me and said, "Actually he should be the one kissing you. He's been in love with you guys for so long." H laughed and said, "He can kiss me if he likes!" I gladly accepted a handshake instead.

I'm sure he doesn't remember that exchange. He must have hundreds upon hundreds of similar collisions with adulating souls each time they head out on the road. But that interaction (after the performance he and his fellow bandmates had put on) was a first-in-a-lifetime for me and would be indelibly etched in my memory.

We met several incredibly friendly people from the audience that night, including Andy Saks (who as it turns out, just happened to live in the next town over from me and would become a lifelong friend). Interestingly, Andy showed up late and the first thing he ever said to us (after *Hello*) was a desperately queried question of "Did they play anything from *Seasons End*?" This has been a running gag with us for 25 years now… Andy, we saw Glen Hansard at The Orpheum last night. He was *so great*! But no, he didn't play anything from *Seasons End*.

Eventually, we called it a night. The Paradise club in Boston contained an entire world that evening. I didn't think that world could get any bigger. Yet there would be many more times seeing this band for me in the years to come in places both near and thousands of miles away. Although at the time, that last fact would seem utterly implausible.

A Note From Julie

So yeah... "Ugh." I had worked a full week, drove 2.5 hours, knew we had a busy wedding weekend ahead and all I wanted to do was get in my PJs, have a glass of wine, order a pizza and watch a movie. And most Fridays, that is what we did. So I remember thinking it was odd that Jeff met me at the door. And his face... He was so excited. At first, I thought it was because I was there. But before I could even say hello, he was talking so fast, all I heard was "Marillion..." and "tonight."

Ugh. So, truth be told. I had no interest in seeing Marillion on this night. If we didn't have the wedding, if I hadn't just driven over two hours, then sure, I would be happy to go. But then I really looked at Jeff. He was like a little kid who wanted to go to the amusement park. It was his Christmas morning and he couldn't get downstairs fast enough to open presents. Sure, I could have let him go by himself. But this was his first chance to see his favorite band. What type of friend would I be if I let him go alone?

As we rode in, he played some from the newest CD. And honestly, the "new" guy looked pretty handsome in the pictures. So worse comes to worst, I look at a cute guy singing for two hours, and I make Jeff's dream come true. I can do that.

The opening song was a Fish song. H came out and commanded the stage and the audience. I thought to myself, "Well, that took balls, this guy has something to him."

And then song after song, I found myself in the palm of h's hand. My heart rose and fell as he spanned octaves in the songs. I couldn't keep my eyes off him. But when I did tear my gaze away, I would look at all the happy faces, screaming every word to every song. The energy was intoxicating. It was pure happiness and bliss. I didn't know exactly what was happening around me, but I knew one thing for sure... I liked being a part of it.

Looking back now, I realize, what was born that night was not only my love for Marillion, the music, the fans. A new connection with Jeff had been created. We shared something that very few could understand. And that something has taken us all over the world. To have moments that we laugh about, cry about and that have shaped our adult lives.

I will always be grateful to Jeff for opening up my heart, my mind and my world. I will also always be grateful to Marillion for creating this family which has brought some of my lifelong friends into my life.

You strong enough to be, Why don't you stand up and say
Give yourself a break, They'll laugh at you anyway
So why don't you stand up and be, Beautiful.

Black, white, red, gold, and brown
We're stuck in this world, Nowhere to go, Turnin' around
What are you so afraid of?, Show us what you're made of
Be yourself and be beautiful, Beautiful

- Lyrics by Steve Hogarth & John Helmer

Chapter 6

Sweet Symmetry

Looking back on that night now, I can see my life divided into two halves. There was the "Before Time." And then there was the time after I had been baptized in the fire of a Marillion live experience.

After that first Marillion concert, Julie was the one who found the *Freaks* list on the internet and told me about it. (I told you she was hooked.) We both joined and kept up to date with every detail of the minutiae discussed. I remember my first post being a plea of never wanting to miss another Marillion show again. From that point on I read every digest and regularly threw my two cents into a heated discussion.

1995 became 1996 and a double live album came out to tide us over. *Made Again* quickly became one of my favorite live albums. And I have to admit I just loved the symmetry. Symmetry? Yes. Allow me to digress.

Fish recorded 4 studio albums with Marillion followed by the double live album, *The Thieving Magpie* before he left the band. H then also recorded 4 studio albums with Marillion, which were followed by a double live album, *Made Again*. Perfect symmetry. Yet it doesn't end there.

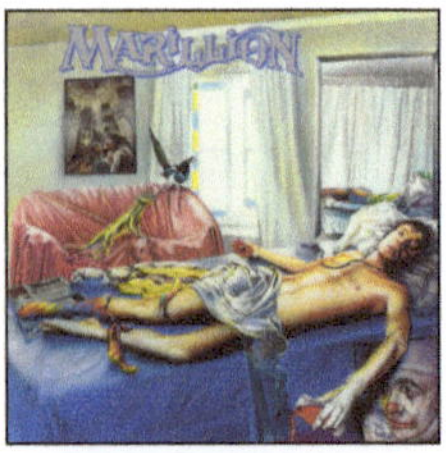

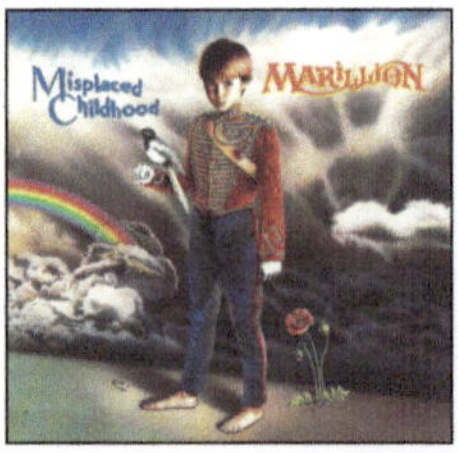

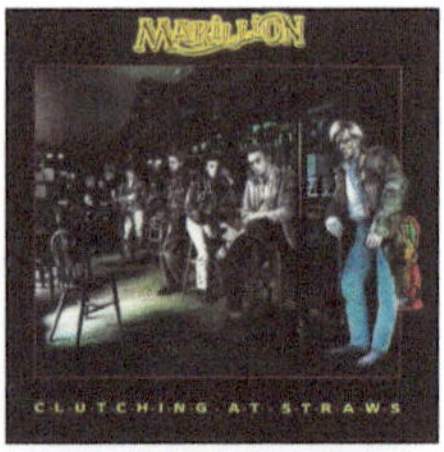

Each singer's third album with the band was a full concept album (*Misplaced Childhood* and *Brave*). And each singer's fourth album with the band (*Clutching at Straws* and *Afraid of Sunlight*) was not a pure concept album but did have a collection of songs loosely tied together by a common theme (the destructive aspects of drinking and fame respectively). And each of those fourth albums was indisputably my personal favorite from those two eras.

Isn't symmetry great? Wait. What? You don't think so? C'mon!... All right. Let's see... How can I explain this better? All right, how about *this* as an example?... Look at these two numerical sequences.

1001001 1001001 1011000 0111011

There is no arguing that the former is just more pleasing to the eye (and the anal-retentive mind of a math nerd) than the latter—and it's not just because it's the binary sequence from Rush's "The Body Electric." It's because it has *symmetry*. The latter is just a jumbled mess. The former with its repeating pattern looks *right*. If you disagree... sorry to say it, but you're just wrong. And yes, I know there were other releases in those two Marillion eras (*Real to Reel*, *B'Sides Themselves*, *Six of One...*) but stop messing with the fun. There's a pattern here...

And yet it didn't end there; they then proceeded to repeat the pattern *again*. Let's recap.

- **Fish-EMI Era:** Four studio albums followed by a double live album – *The Thieving Magpie.*
- **Hogarth-EMI Era:** Four studio albums followed by a double live album – *Made Again.*
- **Hogarth-Indie Era:** Four studio albums followed by a double live album – *Anorak in the UK.*

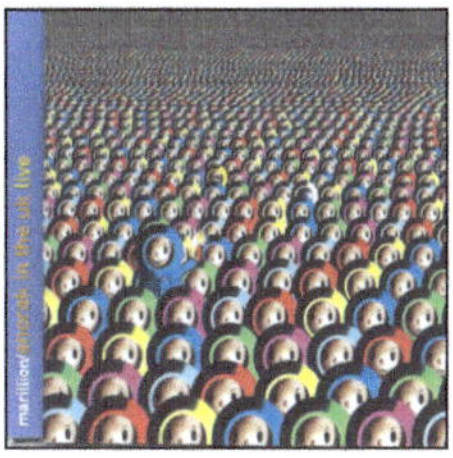

This is fun, isn't it? Was there a fourth sequence? No. They screwed it all up with the double studio album, *Marbles* released in 2004. (Probably my favorite album from the entire "Post-EMI" era.) They then followed that up with the double live album *Marbles Live. C'mon, Marillion. Get your priorities straight.*

Did we get sidetracked again? All right. Back to Mark. This is the last time. I promise. But right after this break.

Rocking Out 1

Pretending I could play Dad's guitar (left). And my first stereo (right) with the requisite early 80's rock on display.

Over the years, I've often gotten so excited about the concerts I've seen, that I would write up reviews and send them out to all my closest friends. I did this in the hope that they would serve as visceral vicarious reenactments of the experiences for those on the receiving end. These missives usually sailed out into the ether and were met with deafening silence. I often wondered why I would keep doing it. Then every once in a while, Mark or Donna or another one of those lifelong friends at the other end of the wireless telegraph would reply back with an excited response. Once I was even told, *don't ever stop sending these*!

Those too-rare words of encouragement made it all worth it. So at this point, you might be thinking, *Cripes, are we going to talk about Marillion for this* entire *book? Most of us have never even* heard *of this band*. OK. OK. I hear you. Let's take a break and delve into a little celebration of the joys of popular music. The contrarian in me always wants to dismiss what becomes overly popular. Sometimes, though, those universal accolades are so rightfully well deserved.

Some bands will lose something from their live performance when they give up the intimacy of the small theaters and clubs for the stadiums and arenas. But others know the meaning of the word *spectacle*. They just know how to put on a massive show.

As we take a moment away from our narrative tale, we will now rock out to Roger Waters' 30th-anniversary performance of Pink Floyd's *The Wall.*

Julie and I would see this same performance again at Fenway Park just a couple of years later. And it was another incredible show with the historic Green Monster serving as the backdrop for Roger's massive physical manifestation of metaphoric isolation. But this first show at The Garden was a mindblower.

Here (in its unadulterated exuberance) is one of my *Hey, Guys, This was Awesome!* re-tellings. (Thanks, "Sent Items" folder.) Kick back and enjoy hearing about music we all know and love.

Roger Waters – The Wall
October 1, 2010

----- Original Message -----

From: Jeffrey Pelletier
Bcc: Everyone
Sent: Sunday, October 3, 2010 11:36 AM
Subject: Roger Waters - The Wall!!!

Well, they dropped some primo seats on the Ticketmaster website at the last minute for Friday night's Roger Waters performance of *The Wall.* So I scooped up a pair of second-row tickets just off the floor only about 5 or 6 rows back from the stage.

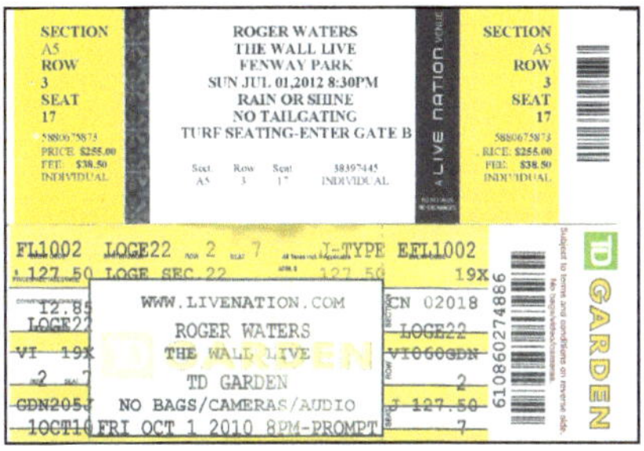

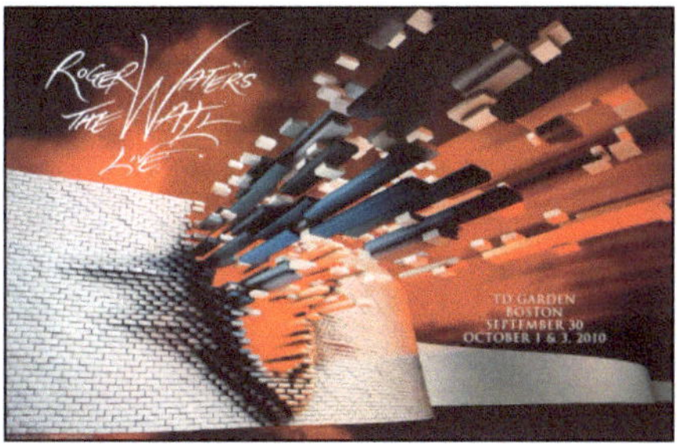

Far enough back to have a perfect view of everything and close enough to actually see the expressions on Roger's face. And raised up just enough to see right over the people on the floor between us and the stage. Sweet! [He actually smiled a lot during the performance. Ha. Who would have thought the bitterest man in the history of rock could smile!]

This was undoubtedly one of the best shows I've ever seen in my life. The music was incredible. But the visual effects were so mind-blowing I won't be able to do them justice.

From the beginning to the very end, it was a non-stop spectacle. This wasn't just a concert; it was truly an event. The pyrotechnics at the beginning of the show during "In The Flesh" shot out over the first few rows (don't know how he got the fire permits...)

Sparks were flying across the stage and flames shot up leaving behind so much smoke it was like the stage had exploded afterward.

The Wall was built brick by brick during the first half of the show. And the Wall itself served as an amazing video screen for front projection and rear projection of video effects.

Pictures of fallen soldiers and commentary to the songs were displayed at various times. During "Mother," the line "Mother should I trust the government?" was followed with a gigantic spray-painted response across the Wall of "*No Fucking Way!*" Ha.

The second half of the show was performed at times *through* the Wall, *on top of* the Wall and even *within* it—with a fully functional living room (with lights and TV) folding out of The Wall for "Nobody's Home."

Projections of The Wall onto the Wall showed the Wall opening and closing in a way that was so realistic you couldn't help but cheer. The original animation was used at times and at other times it was completely updated with all new effects. The bomber sequence during "Goodbye Blue Sky" was stunning.

Massive 20- to 25-foot animatronic puppets of the Teacher and Mother leered out over the audience. A group of school kids ran on stage to sing the chorus to "Another Brick in Wall" (Part 2, of course). The Pink Floyd pig made an appearance flying over the audience close enough to touch.

But at the end of "The Trial," when the entire audience is screaming along to "Tear Down The Wall!... *Tear* Down the Wall!!!" with fists in the air... The Wall didn't just come down a few rows of bricks at a time like on the DVD. The first few rows dropped... Then the rest of the Wall shook and shuddered and then came crashing forward with massive bricks actually falling off the stage in front of the audience. It was un-freakin'-believable.

And the music was awesome!!! Every song was fantastic, with the guitar solo of "Comfortably Numb" played from the very top of the Wall being the highlight of the night. "Run Like Hell," "Young Lust," "Hey You," all amazing. And the way Roger could incite the crowd to a frenzy during all the pseudo-fascist play-acting was just incredible. You couldn't help but get swept up into the excitement...

It's not too late to see the show. Tickets are still available for tonight (Sunday)... I'm just now recovering, and Julie (who wasn't even that excited about it at first) was blown away, saying it was probably the best show she's ever seen as well.

Don't Miss out!!!! :)

So if ya thought ya might like to, go to the show; to feel the warm thrill of confusion, that space cadet glow... Then don't miss out! :)

Woohoo!!!

That was truly AWESOME!!!!

Story 3

I am a reader. I bring a book with me everywhere I go. And I do mean *everywhere.* I went to see the first *Lilith Fair* concert at Great Woods in 1997 because I was a big fan of Sarah McLachlan and Paula Cole. But when Fiona Apple took the stage, I pulled a book out from under my seat and started reading, since it was still light out. I heard some bemused chuckles from those seated around me, and I didn't care.

I could fill many pages talking about the stories of some of my favorite authors—including John Irving, Orson Scott Card, Douglas Adams, Tom Clancy, Tom Perrotta and Nick Hornby just to name a few. I won't do that though. You're welcome.

My parents told me that as a young child I was always reading. I honestly don't remember this or any of those books from my early years. It felt to me like my love of reading came to me in my adolescence. As I looked back and tried to remember some of the earliest books I fell in love with, there was one that stood out. It was a book that I didn't finish for a long time because of my inherent laziness as a kid, but it came back to me in re-released form many years later. And so here is my Mockingbird tale, told at the time of that rediscovery.

Age 15, Grade 9, 1982
"Mockingbird"

It was the last week of school at West Junior High in Andover and I was in a *great* mood. Not just for the obvious reasons, but also because of something I never would have expected... a *book.* A beautifully written, wonderful book that taught me how each person has within them the ability to touch another person's life. Or to quote

Opus the Penguin, "reminds us of the fragile thread of dignity that unites the human tapestry."

The book? "To Kill A Mockingbird."

The public schooling system would later introduce me to some really great classics that set my imagination alight, ("Great Expectations," "My Antonia," "Macbeth"...), and it would also force me to trudge through a lot of stuff that just struck me as grandiose literary crap or endless drudgery, ("The Old Man and the Sea," "Beowulf"...) But no book that I ever *had* to read would reach into my very being the way this one would. I truly loved discovering this story.

So anyway, it was the last week of school and I only had about 25 pages left in "Mockingbird." I was tempted to stay up late and finish it the night before, but I'm a slow reader (I like to speak every word in my head and hear the rhythm of the syllables and the flow of the words). So, I was saving the end for tonight even though I was *supposed* to have finished it already.

I was unsuccessfully trying to think up a way to initiate a conversation with Nicole Durant sitting next to me. Nicole was cute and nice in a non-stuck-up way, despite her popularity. It's not like I thought I had a chance with her, but this was the smart kids' class. And, in comparison to all the nerds and social rejects in this room, I had to look like an acceptable option, right? (That's not to say I *wasn't* a nerd or social reject too... *just in comparison*).

But wait a minute... What's the teacher doing? She's calling each of us up front to turn in our books. No. *Not yet.* I'm not finished! There are still three days left of school. OK, no problem. I'll just ask her to keep it another day. *In front of everybody, Jeff? They'll all laugh at you*. Well, I don't care. It's worth it.

Mrs. Harrison eventually called my name and I walked up to her desk. Just as I was about to ask her for another day, I realized something terrible. *We already took the test! If I ask her to keep the book, she'll know I haven't finished it yet. What do I do?*

She held her left hand out across her desk, waiting with her right to place a checkmark next to my name and move on to the next student on her alphabetical list, Judy Potvin. I started to give her the book but hesitated. She raised her head and our eyes met.

She knows. Who cares? Ask her! No, I can't. She'll give me a lower grade. But you have to find out what happens! Will Scout and Jem ever see Boo Radley? Will Dill come back again next summer? And Atticus? Oh, Atticus! What's going to happen to him with that awful Mr. Ewell making all those threats?

I gave her the book and returned to my little desk. I didn't talk to Nicole or anyone else for the rest of the class. I felt like I had just lost my best friend.

→ Fast Forward Thirteen Years →

I never found out how "Mockingbird" ended, not for 13 years, that is. I never checked it out of the library nor picked up the paperback at a bookstore. It just didn't seem the time was right. Until now.

Last month, I was in Barnes and Noble and what did I see? A beautifully-bound hardcover 30th-anniversary edition of "To Kill a Mockingbird" with a new foreword by the author. Actually the "new Foreword" is just a letter that Harper Lee wrote to the publishers begging them not to put a new intro in the anniversary edition of Mockingbird – Ha. Hilarious. (Who knew book publishers had a sense of humor.)

I eagerly bought the sucker and did something I usually never do. I stopped reading the book I was halfway through to spend the next week back in Scout's Maycomb County. What a wonderful nostalgic trip back to my childhood. Now the only question is, do I pick up where I left off in the story way back then, or start from scratch? (*I started from scratch, and it was worth it.*)

"Scout, Stand Up! Your *father's* passin'."

Story 4

Human memory is incredibly unreliable. We all obstinately insist that these details we lived through *have* to be true. We remember them so *clearly*. And so often they simply are not. I learned of a fascinating theory recently about how our memory might in fact work. It's been suggested that each time we remember an event, we're not actually remembering the event itself. We're remembering the *last time* we *remembered* the event.

Do you realize the implications of what that means if it's in fact true? Every time we recall something from our past, we're re-writing over the memory file in our brains. We are erasing the original data and filling in what we believe is true. We could talk about this for hours—especially the implications of this on eyewitness testimony. But why are we discussing it now?

I wrote this story exactly as I remember it. I know my memory of the events has to be distorted by the years. The memory file in my brain has either been re-written many times or simply suffered some non-insignificant data loss by now. Additionally, as you read this, you're bound to assume that there's a lot of exaggeration thrown in here for comedic effect. All I can tell you is, this is truly how I remember it happening to the best of my recollection. Yes. I was a complete idiot. With a girl. Does that surprise you? It shouldn't.

At our 25th reunion, in 2010, I had the opportunity to speak to Kim Sullivan, the first girl I ever asked out. I thanked her for being the first girl who ever showed interest in me as a boy. And I also apologized for the embarrassingly dramatic way I ended up doing it. She laughed and said she didn't remember it being so crazy as I described it. But here's how it went down in my memory. This one's for you, Kim!

Age 12, Grade 6, 1979
"Admiring Secretly"

In sixth grade, Tim and Andy were my best friends. We did everything together. And here's the thing. Andy and Tim had girlfriends, and I didn't yet. It wasn't like I *had* to have a girlfriend just because they did. Though my hormones were starting to kick in. And every time I saw a pretty girl on TV, I was imagining myself kissing her. And I actually had my eye on a particular someone. So it seemed that my time had arrived. I was actually going to *Ask Someone Out.*

Of course, *Going Out* didn't mean that we actually *went* anywhere; it just meant that we had agreed to share a nebulous state of being *thought of* as "together," without ever really spending any quality time talking and getting to know each other. It's what I imagined *getting pinned* was back in the 50's, minus the actual pin.

Was she the prettiest girl in school? Maybe not, but she sure seemed cute to me. Was she the most popular girl? Again, probably not, but she seemed pretty cool—not that I had actually *talked* to her before or anything. But she could kick a kickball hard enough to get on base and she was always laughing with her friends, so she must have a sense of humor too.

So, I was at Tim's house one night and we were talking about how I was going to ask Kim out. I told him it was just too much. There's no way I could summon the courage. I was too nervous to call her and I didn't know how to approach her at school. Tim suggested I write her a note.

This seemed like the perfect solution. Given the time, I could mold and shape the perfect words to win her affection. So this is exactly what I did. Yet, when it came down to signing my name, I again didn't have the courage. I chickened out. I decided that I'd send the note to her anonymously. Then I could worry later about approaching her and asking the actual question. And so, in a misguided attempt to be funny, I made the fatal mistake of signing the note, "Your Secret Admirer."

I thought this would be kind of stupidly cute and I'd break the ice right after she read the note and this would give us something to laugh about. I was waiting for verification that my note was received before I would steel myself and do the deed. But before I could be sure she got it, word had made itself known. And that word traveled *lightning*-fast. Every single student in Andover West Elementary Grade 6 knew that Kim had a secret admirer and they couldn't wait to find out who it was.

I couldn't do it. I couldn't reveal myself as the bashful admirer. Why did I do this? How was I going to fix it? Instead, I sent her another anonymous note the next day and hoped that this time I'd be able to talk to her without the attention of the entire school. (Obviously, I was no longer thinking clearly by this point.) Boy, was I wrong. The fervor and excitement were even greater than yesterday. It was all everyone was talking about.

The third day was no different and I knew it couldn't go on much longer before someone blew my cover. Before first recess, I slipped a note in her locker that said, "Kim, I'd like to talk to you at recess. Jeff." As the note fell through one of the three horizontal slots into her locker, an eerie calm fell over my body. There was no turning back now. The Rubicon had been crossed.

Next thing I knew, recess was upon us. First recess was an indoor recess, so everybody was hanging out in the classroom and in the hallway. I was no longer feeling calm. I was pacing back and forth with my buddies in the classroom, stealing glances down the hallway, waiting for Kim to return from her locker with the evidence of my folly. A small group of my friends knew it was me, and we huddled together waiting.

Finally, Kim and her best friend (also named Kim) came into view. They were surrounded by people whispering in each other's ears. Back in class, I steeled my nerves and tried to convince myself that I would survive the next five minutes. The noise level dramatically increased, so I knew she had arrived. With a few last words of encouragement from my friends, I began the "long walk."

As I caught a glimpse of the two Kims coming in from the hallway, I noticed my note in my Kim's hand and a smile on her face. *Hey, this might actually all work out,* I thought. And then Austin Wang grabbed me by the arm and yelled, "Go get her, Jeff!" With all the force he could muster, he swung me around in a circle twice and then propelled me out into the hallway. I flew by the two Kims, dizzy and completely off-balance.

My back slammed into the end of a table and it almost snapped me in half like a twig. This had the double effect of both sending a bolt of searing pain up my spine and of knocking the breath completely out of me. I collapsed onto my butt on the floor and looked up into what seemed like a hundred stone-cold silent eyes; many with mouths hanging agape in astonishment.

I looked up at the Kims through my fear-sweat-pain and humiliation-clouded eyes and saw her face contort with conflicting reactions. It looked like half of her wanted to laugh out loud at the absurdity of what had just happened and the other half was actually concerned for my well-being.

All was still and silent as everyone stared at me sitting on the floor in front of the two Kims, so I figured I might as well get it over with. It's not like I could embarrass myself anymore at this point. I had pretty much reached that summit. With all the semblance of casualty I could summon, I said, "So, Kim... you wanna go out with me?"

Thank all the Gods from all the pagan rituals that ever existed... she smiled and said "Sure."

Success! I stood up and scurried away into the crowd as fast as I could, fleeing the scene like a criminal. Behind me, I thought I heard a cheer arise from the masses, but maybe they all were just laughing, finally releasing the tension from the "romantic" scene they had just witnessed.

Someday I may top that one for "great romantic gestures," but I haven't come close since.

And now, back to our main storyline...

The Announcement

As 1997 approached, details started to emerge about the next album. Marillion often joke that they write new material on pace with the movement of glaciers. And to a diehard fan, it sure can seem that way. But at long last, we finally had news from the band that the album was done and it would be called *This Strange Engine.*

Tour dates were announced for Europe. And then on the fateful day of January 27, 1997, Mark Kelly posted to the Freaks List that they would unfortunately not be able to tour the United States or North America at all. As it turned out, we learned that previous North American tours had always lost money for the band.

The previous tour had lost £30,000. At the time of its writing, Pete Trewavas further explained to Jon Collins for his excellent 2002 biography *Marillion/Separated Out,* "Over the years, we'd spent so much money and time and we weren't really getting anywhere. There's still a club circuit in the US that even the bigger bands still follow. It's a strange thing for European bands to come to terms with."

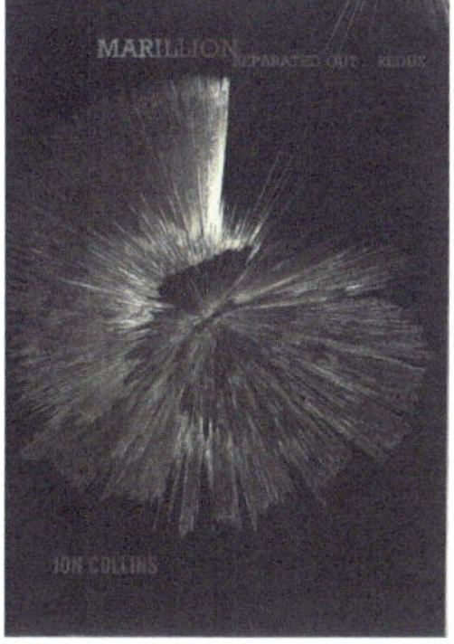

With Jon's kind permission, we see the great "Separated Out" cover (left) designed by Darren Newitt, as well as "Separated Out... Redux" (right)—his updated 2012 version.

Delivering the disappointing news, Mark added the following to the Freaks List:

"Hello to all the best Freaks...

We love touring. It is just too expensive to tour some places! I'm not saying it's impossible to tour everywhere, but it is expensive to put on a Marillion tour... We are not prepared to put on half a show with second-rate equipment and a crap sound system, just so we can get to the west coast and not lose our shirts!"

Promoters wouldn't allow them to charge more per ticket, because they didn't think people would show up. And even with merchandise sales (the prime profit source for touring bands), they would not be able to make a profit. Imagine working for several long grinding months on the road to later learn that not only did you not make anything for all your effort, but you end up owing way more than you took in. And so, sadly, a Marillion tour across the Atlantic was not going to happen.

Needless to say, I was devastated. I had finally seen the band I privately loved on their last tour. I had finally successfully "converted" a new fan. I had found this community that had kept my incessant desire for *more, more,* ***more*** smoldering with the heat of a thousand suns. For the first time, I had a glimpse into the creation of a new album and saw its arrival approach as updates were posted to the *Freaks* list. And now I learned that I wasn't going to have the opportunity to experience that live-performance *high* again.

Chapter 8

Generosity or Selfishness?

I have to admit, that what motivated me to suggest what was kind of an out-of-the-box idea for the time was not generosity. I had had my taste, and I needed another fix. I was an addict. My mind was flailing about for any way to make that happen. So, I consider what I dreamed up an act of pure selfishness. (*I want an Oompa-Loompa* ***now****, Daddy!*)

Remember the friggin' Eagles? Well, now we will see why we talked about them a few chapters ago. And just maybe, now we can better understand why Jeff Lebowski said, "Man, I *hate* the fucking Eagles, *man*."

My one and only Marillion concert had cost me a whopping 12 bucks (plus fees—now up to $2.50). And The Eagles had charged $100 a ticket and *more*. Surely Marillion could charge more, right? At the time, the answer was a resounding "*No*." Apparently, they had tried to convince the promoters that they knew their fans. They would pay more. (I knew I would pay more.) But the promoters weren't going to have any of it.

The Realization: Well, why not just cut out the middlemen?

Why not take the difference between what the promoters would *let them* charge and what we (the fans) would be *willing* to pay… and just give that directly to the band?

OK. Let's think about this. We're all nuts about this band. But realistically, even though all of us would *want* to pay more, some of us probably can't afford to. But if it meant no tour at all, would *half of us here* be willing to actually take money out of our wallets and just give it away, sight unseen?

There were close to 2,000 people on the *Freaks* list at the time. A lot of us probably couldn't pay an extra $100 each like The Eagles fans. But what if just half of us came up with an extra 50 bucks each?

Let's do the math. $50 x (0.5 x 2000) = **$50,000.**

And what if only a quarter of us contributed? $50 x (0.25 x 2000) = **$25,000.**

You know what? That could actually *work*. $50K would certainly cover whatever they would lose on a tour. But even $25K might be enough to make it feasible. And I think just *one out of four of us* is a realistically achievable goal. I mean, we are all *nuts* about this band.

Consequently, I sat down at my keyboard to simply vent my frustrations. I wrote a post to the *Freaks* list. I touched on a few different topics in the same post (as was my habit). But I also decided to include my idea. I didn't in my wildest dreams think it would amount to anything in the real world. I just wanted to fantasize for a moment that I could change this heartbreaking turn of events.

I still have that original email all these years later, believe it or not. Here's the relevant section of what I posted. This email, years later, would be quoted in multiple books about Marillion as well as being used in lawsuits regarding crowdfunding intellectual property.

Date: Mon, 27 Jan 1997 16:03 -0500 (EST)
From: PELLETJ@cliffy.polaroid.com
To: freaks@ax.com
Subject: EMi reply, USA touring
Message-ID:

> Here's a radical idea. What if all the US freaks donated to a "Marillion Tour the USA" fund? I'd gladly throw $50 in the hat if that guaranteed a chance to see them in several US locations. (And I would hope that preferential seating would be made available to contributors, in case they venture beyond the clubs to some smaller theaters.) What do you think? There's several hundred US Marillion fanatics on this list alone, right? I know it wouldn't cover all the costs but it might help...

All for now,
Jeff Pelletier
pelletj@polaroid.com

Chapter 9

An Unexpected Call

I didn't have a home computer back in the 90's. I used my work computer for all my personal e-mail activity. Corporate time theft, you ask a little accusatorially? Maybe a little. But I was averaging at least 50 hours a week at my job trying to impress the bosses and accelerate my trajectory to financial success. Therefore, I didn't see a problem with a bit of time spent on the job for personal distraction.

The next day, I got sucked into actual work in the laser lab (as I often did back then), and hadn't been thinking at all about what I had written the day before. I hadn't checked my e-mail in the morning and worked straight through lunch. It was only in the afternoon that I walked into my office and slumped into my swivel chair just as the phone started to ring.

I picked up and said, "Jeff Pelletier" in that matter-of-fact way that said *I'm at work and saving syllables by just identifying myself as the person you just called.* I then hear Julie's voice, saying, "You're a genius!"

Huh? I asked what the heck she was talking about and she said, "Haven't you seen the *Freaks* list? Pledges are pouring in." I opened my e-mail and my inbox had about 10 times the usual number of *Freaks* digests. I started scrolling through them and what did I see? Over and over again, people had said they would definitely contribute to a Marillion Tour Fund. Pledges came not just from those of us here in North America, but from all around the world.

After several digests of pledges were filled, another fan named Jeff (Woods), volunteered to set up the fund and take the payments. Through his private business, he had the ability to accept payments by credit card. This was a big deal back then (again, before the days of PayPal and Venmo and easy, free instant payments to individuals).

More pledges came in and before we knew it, Mark Kelly posted saying he had told the band what we wanted to do. They all thought we were crazy but they gave us the go-ahead to give it a shot.

Chapter 10

The Truly Generous Ones

Here's one of the coolest things about the Tour Fund. Before the pledges had reached that tipping point level of *Hey, enough of us are really doing this that it could actually happen…* People who didn't have a chance of seeing any of the shows were pledging money just to see it succeed.

Think about that. They wouldn't get to see a show and wouldn't get anything in return for their money. Yet they still gave. How amazing is that?

Shortly thereafter, Mark Kelly further announced that they (Marillion) wouldn't feel comfortable taking our money without giving something back in return. If the Tour Fund was a success and they did in fact tour North America, they pledged to us that they would record the concerts and make a CD of the best show to give to everyone who donated at least 25 dollars to the fund as a Thank You gift. And what did that do? It created a second wave of pledges that rolled in. Now people were actually sending their money to Mr. Jeff W and the special account that he had set up. Pledges had become reality.

You see, so many Marillion fans were "completists." A completist is a collector that has to own *everything* ever released by the band. All those rare singles and picture discs and EPs were *must-haves* to a completist. And this ultra-limited-edition CD that would go out to all the Tour Fund donors would become the most coveted rarity in a Marillion collection.

The Tour Fund would not have happened without the instant communication and connectivity made possible by the internet. A call to action by regular mail would not have received the instant and eager response that manifested on the Freaks list in that single day. It was only the sheer volume of positive feedback (and the speed at which it came) that convinced Mark Kelly and the rest of the band to let us give this crazy idea a try.

Chapter 11

A Final Hurdle Leaped

Now let's jump forward only a couple of months. Once the Tour Fund had collected about $30,000 of real money in hand, the idea had been deemed a success and tour dates had been booked. However, storm clouds were brewing on another front with regard to the new album.

Marillion had previously recorded 8 studio albums (and other releases) with the industry behemoth, EMI. With the end of their contractual obligation to a major label, they had looked at getting themselves a better deal with a smaller label. So, with this first post-EMI album, they signed with Castle Records.

At first, there wasn't even a plan to release *This Strange Engine* in the United States, much less tour there. However, a distribution deal was later made with another indie label named Red Ant—a record label that was distributed by Polygram/Island Records. Initially, Red Ant was also against the idea of the Tour Fund. Yet, begrudgingly they agreed to the band's request that they match the Tour Fund's $30,000, and a small number of tour dates in North America were scheduled.

But guess what? Red Ant just declared that they were going bankrupt.

This put everything, including our fan-funded tour in jeopardy. Now if we were going to see our grand project come to fruition, we were going to have to be the *sole* underwriters of an entire international tour. An update was sent to the *Freaks* list that we needed to meet a minimum threshold if this was going to be feasible. I believe somewhere in the range of $30,000 to $40,000 had been collected at that point. But if we didn't reach $50,000, we were going to have to cancel the tour and return everyone's money.

What followed was the third and final wave of donations. Did we reach $50,000? Indeed we did. We actually raised over **$60,000**. And thus, for the first time in the history of music, an international tour of a major label band was funded *without a single cent from the music industry*. Pretty cool, right?

Red Ant never released *This Strange Engine* in the United States. Eventually, a distribution deal was reached with Eagle Rock/Velvel Records. They got the CDs onto the record shelves here at long last. However, most of the tour had already taken place before that even happened. I distinctly remember thinking that if I hadn't special ordered an imported UK version of the CD (a very new thing for me at the time), I wouldn't even be familiar with the new music that was being featured on this tour.

When asked about all of this by Prog magazine, Mark recalls the following:

"People used to communicate via mailing lists. And someone asked the question, would we be touring in the states and I said, 'No, we've just left EMI, and whenever we toured America, we always lost money, and the record company footed the bill.' Well, the next question was how much would you lose? And I said, 'Well, maybe fifty or sixty thousand dollars, thinking that would put an end to the conversation.

Then this guy said, 'Well, why don't we raise the money?' And then some other guy said, 'Yeah, I can open a bank account and put some money into an escrow account, and if the tour doesn't happen, I'll just give the money back to everybody.' I was slightly concerned, but I was thinking that not much would happen. They raised about $20,000 in the first week, then I had to go and tell the rest of the band..."

Although there are a lot of other quotes where he mentions me by name, I love that Mark refers to me and Jeff W. as "this guy" and "some other guy" in this one.

Also, I swear I didn't plan it, but you have to appreciate the coincidence of the bankruptcy being discussed in "Chapter 11" of this book.

You don't need money to be rich anyhow
Spending yourself is what it's all about
... Spend yourself and get rich right now.

- Lyrics by Steve Hogarth

Chapter 12

OK. But Who the Hell Is This Band "Marillion"?

As someone who entered high school in the early 80's, I thought music videos on television were the coolest thing to hit the airwaves. When MTV first debuted, a lot of us didn't have access to the channel because we had to pay extra for it. But in those days, you could tune in to the channel and the picture would be scrambled, but the audio still came through clearly.

I laugh to think about the hours we would spend sitting in front of the TV watching that channel—sometimes in front of a wavy screen, not even *seeing* what they were broadcasting. Ha. (See, I actually *did* laugh). In the early days, they didn't even have that many popular bands on MTV. It was mostly avant-garde videos of European bands. I think back to all those *luftballons* and *dances of safety* we sat through. Still, we waited and waited and leaped forward with excitement when one of our favorites came on.

The video for the song "Kayleigh" was in heavy rotation on MTV in the states when 1985 rolled around. (Oddly I don't remember ever hearing it on the radio, just seeing it on MTV). I recall liking the "Kayleigh" video with its unique sound and its nostalgic and introspective "lost love" theme. Yet, the song seemed to end somewhat abruptly and I didn't explore Marillion further at that time. It wasn't until the fall of 1987 in the first semester of my junior year at college that Marillion would come back to me with permanence.

There were several of us in the dorms at U. Lowell that were on this wonderful voyage of musical discovery. We dug and clawed into the past 20 years of music until our fingers were raw. We ventured into the realms of early Genesis, Yes, and Floyd with a sense of amazement and joy—as well as some more eclectic popular contemporaries of the time like Steely Dan, Peter Gabriel & Joe Jackson. How had we wasted so much time on such generic music before?

Ed Murphy (seen at right displaying his love of Marillion's original singer, Fish, on Dan Peduzzi's dorm room door) was a good friend and fellow music lover across the hall. One day he came to me with Marillion's *Misplaced Childhood* album in hand. He said he had something intriguing for me to check out. (Thank you, Ed! You had no idea what you started with that simple act.)

"Kayleigh" immediately relit the spark from a couple of years before, and I soon found out why the video ending seemed so abrupt. It was actually one part of a long *suite* of songs that were meant to be played as one. The song was infinitely better in the context of the entire album. I played that CD several times, and with each successive play, I was drawn in a little bit more. I had to venture further.

Clutching at Straws was the newest CD by this unexpected band, Marillion. As soon as I heard the "Warm Wet Circle" trilogy for the very first time I was captivated, enthralled, *mesmerized* and left in a state of bewildered awe. This was literally unlike anything I'd ever heard. Such poetic phrasings and visceral images weaved into these musically complex tapestries in a way that was completely accessible to anyone with a sense of adventure and a mind open enough to appreciate it.

Those first four Marillion CDs (from that initial era of the band with original lead singer, Fish) immediately took the prime slots in my listening rotation along with Roger Waters' *Radio KAOS* album which also hit me at the same time. Marillion however was mostly a private passion. Most of my musical comrades weren't embracing this incredible band the way I had—something which still confounds me to this day. There were only a couple of us who loved Marillion. The rest soon tired of hearing them (and hearing *of* them) all the time with my endless proselytizing.

They were still passionate about music—just not this band. One college memory permanently etched onto my brain is of waking in the pre-dawn hours to the bright room lights *searing* through my eyelids. My roommate Dan Riemer had scrambled out of bed when he heard his favorite tune ("Midnight Special" by CCR), had come on the radio. (Yes, we slept with the music on in the background. Why wouldn't you? Sometimes the songs actually enter your dreams.)

As I squinted in groggy confusion to see what was going on, I was presented with the image of Dan's hairy back as he shook his booty with revelrous abandon in nothing but his tighty-whities with my headphones clamped over his ears at full volume... "*Let the Midnight Spe-ecial.... Shine a light on me...*"

By the time I discovered Marillion in '87, all the studio work that Fish would produce with them had come to an end, though it would take some time for me to find that fact out. In those days, my favorite hobby was to go to the local music store once a week and scan through the CD racks of all those bands, looking for that unexpected new surprise.

The M section was always where my anticipation peaked. I had found a few imported gems before, a live CD, a collection of B-sides... but a new studio CD was a long time coming. It wasn't until 1989 that I was finally rewarded for my vigilance.

Flip, flip, flip the CDs in the rack... Flip, flip, flip... then... *small gasp of surprise! Seasons End* was there along with the other four Marillion CDs. After a short, appreciative study of the cover art, a quick look at the copyright date (verifying that it was in fact *new*) and an inspection of the song titles, I bought it on the spot.

I had a couple of friends who were still following the band and one had recently heard that Fish had left the band. I didn't think the band could continue without him (as the sole lyric writer, in addition to that singular voice) but here was evidence to the contrary. (What a whopper of an unnecessary worry that would turn out to be.) I drove back to our grad school apartment as fast as I could.

Anyone remember the days before car CD players, when you couldn't play your new CDs until you got home? After a few listens at home, I would then pull out a blank Maxell cassette and make a tape of the CD to play in the car. Much thought would go into what would go on the flip side of the 90-minute medium.

What a relief when Maxell realized how many CDs were over 45 minutes each and they started making 100-minute tapes. Yet sometimes you still had to sacrifice a song to make it fit within that 50 minutes (51 and change actually, if you were paying attention). I could usually find a track I could live without, but sometimes it killed me to pick which one wouldn't make the cut.

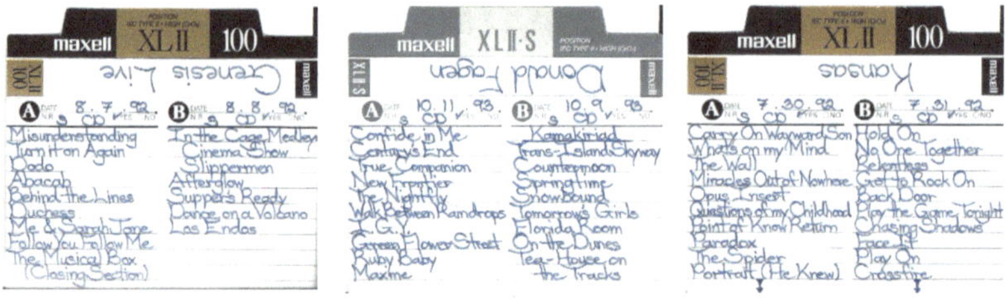

I have horrible cursive penmanship. Yet when it was time to fill out the Maxell insert with the song list, I would ever-so-slowly carefully print them out. This signature style was appreciated enough that several friends usually requested I do the same for their tapes.

I was quite relieved to find that I really liked *Seasons End.* The new songs didn't seem to have Fish's lyrical complexity and the voice wasn't anything like Fish's near-falsetto... but it was good, and *strong*. It belonged to a new singer named Steve Hogarth. The music was excellent. That same evocative character was still emanating from these songs even with this new updated sound.

Two days later I came home from class to find a note on my dry-erase message board outside our apartment door that simply said, "Who the hell is Steve Hogarth???" Obviously one of the few Marillion appreciators I knew had gone through the same experience with this new CD.

In time, *Seasons End* proved to be as strong a first album as one could hope for from the new lyricist and vocalist in my favorite band. Yet their follow-up album *Holidays in Eden* left me feeling somewhat disappointed *at first* with the direction the band was taking. The US version opened with two quite poppy songs in "Cover my Eyes" and "No One Can." Don't get me wrong. They were *very good* pop songs. Yet four-minute fun songs weren't what I had come to expect from the band that had previously taken me on album-long journeys.

My initial underwhelmed reaction to these songs kept me from fully appreciating a majestic track like "Splintering Heart" (which was the opening track on the UK version of the CD) until months later. "Splintering Heart" is one of those bombastic, epic roller-coaster rides of a song that Marillion do best. It culminates in a cathartic vocal screaming climax that sends shivers through the listener.

Years later, while attending a John Wesley show in New York, we heard him talk about the first night he started working as a roadie for Marillion—not even knowing that Steve Hogarth was the lead singer. John was moving and setting up equipment as the band was running through their soundcheck.

*They break into "Splintering Heart" and he says to himself, "Jesus **Christ**! What the fuck is **this**?" And by the time h gets through singing "100 Nights" and does "that thing at the end," John tells us he's thinking, "There's no way he's going to pull this off every night. There's no one who can pull this off every night... And 260 shows later he pulled it off... every night."*

So you save up your tears, for the moments alone
'Til the splinters you gather, leave you glass-hard and numb
And the same sun is shining, on the old and the young
On the saints and the sinners, on the weak and the strong.

And there's a burning and freezing, and a cross for a kiss
So she learns to stop dreaming, and you know how it is.
With these fragments of love, and this splintering heart.
With the fragments, and this splintering heart.

- Lyrics by Steve Hogarth

The Marillion led by Mr. Hogarth completely won me over for good with *Brave* (1994) and *Afraid of Sunlight* (1995). To this day, after listening to these albums about 1000 times each (without exaggeration), I can still be brought to the verge of tears at the sheer emotionality packed into these two masterpieces.

And since that time, they've released nine more studio albums each containing tracks worthy of the never-ending debate of which are their finest songs. And so, this is the band Marillion to me. That band is the reason you're reading this book if you're a fan. Or conversely, it's the band that you're wondering why they, of all artists, were the ones to prompt the first ever internet-based crowdfunding campaign if you're not.

Rocking Out 2

That first review was fun, wasn't it? Are you up for another? Great. Then let's divert from Marillion once more for something just a little bit more well-known. And if not, feel free to skip forward a few pages. None of this is required reading.

This time we turn to possibly the biggest rock band of current times: U2. I've been lucky enough to see every U2 tour (with the exception of *Pop*) since "Zootopia" rocked the old Foxboro Stadium on the *Achtung Baby* tour. Each one of those tours is arguably one of the greatest shows I've seen. (Again, some bands just know how to do *spectacle*. And U2 definitely fits that bill.)

This time, however, I was presented with a unique opportunity. Previously, Julie and I had seen the *Elevation* tour at the Boston Garden in the Summer of 2001. Then in October 2001, when I won tickets to the Providence, Rhode Island, show on the radio, Julie didn't have it in her to make the trip. I asked lifelong friend Melissa Zampitella (previously Melissa Witty) and she jumped at the chance. (Yes, this is the same Melissa I mentioned from the Eagles concert back at the beginning of the book. She rocks with the best of them.)

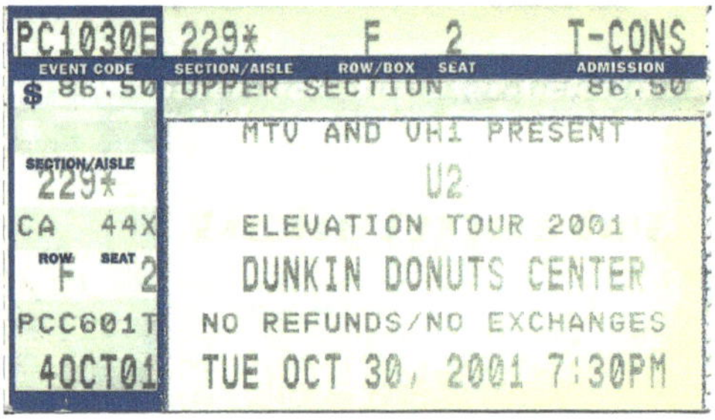

This was the 2nd leg of that *Elevation* tour that included the post-9/11, incredibly moving listing of every name lost on those fateful flights. The names scrolled by on giant vertical screens on either side of the stage as the band played "One" to gorgeous effect, followed by the new song "Walk On" in perfect, poignant juxtaposition.

Melissa had long considered that one of her favorite concerts of all time. Because of that, I was the first person she thought of when she had the opportunity to return the favor. Initially, Melissa offered to treat Julie and me to a free U2 concert (which was great in itself). Quickly to follow though was an offer to actually *work for the band* in a very small way. We leaped at the chance... Once again, here is the original email sent out in sheer exhilaration once the experience finally came to a close.

U2 – Songs of Innocence
July 10 – 15, 2015

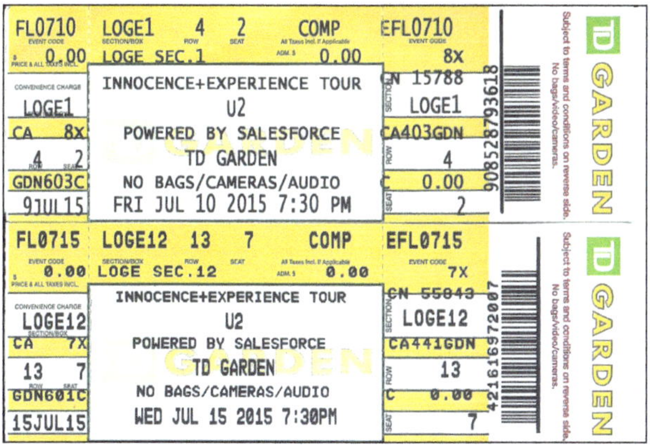

----- Original Message -----
From: Jeffrey Pelletier
Sent: Friday, July 17, 2015 2:29 PM
Bcc: Everyone
Subject: U2!!!!!!!!!!!!!!!!!!

What a week it's been! It started last week when my dear friend Melissa said she could get some free corporate seats to the U2 show and wanted to take me and Julie – to thank me for when I took her to a free U2 show back in 2001. Fantastic!

It turns out Bono is an investor in her company *Vride* and they are sponsoring the whole tour to promote their new *Ride* app. People who use the app to carpool to the show get free parking and early VIP admission and are entered into a raffle for back-stage access.

Well, another call from Melissa soon followed and I learn that they need people to help usher in the VIP "early entry" people.

Julie and I sign up, and the next thing we know we have free premium seats to Friday, Saturday ***and*** Wednesday's shows, with free parking and even get paid for our time. Woohoo! ($300 face value per ticket x 3 nights is **$1,800** worth of tickets plus another $100 in free parking, not to mention some extra cash on the side for our "work." Nice!)

From 5 to 6 PM each night, Julie and I are stationed on the 2nd and 3rd levels of the Garden directing about 40 or 50 lucky people up to the concourse where the raffle takes place. While we are doing this, we get to hear U2 running through their sound-check. It's as fun as a job can get. Then after an hour, our duties are done. We walk over to the North End for dinner and a drink (paid for by the hour of work we just did) before the show each night and then it's back for the concert. Easiest job ever! And now on to the shows...

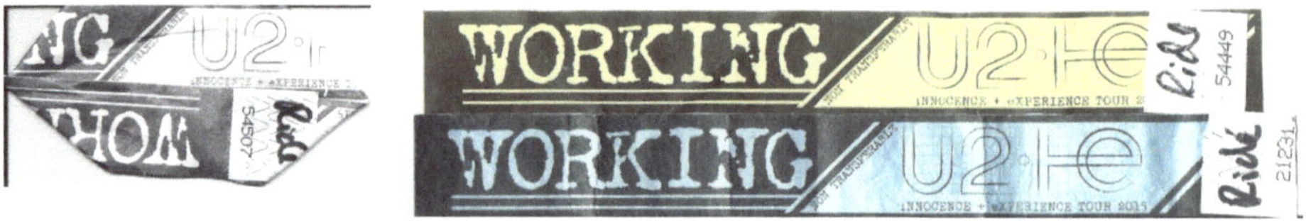

U2 "Working Bracelets" – Wrist up as you enter: "Working for the band. Working for the band..."

The stage design that U2 has for this tour is one of the best I've ever seen. As with Peter Gabriel's *Secret World* tour, the main stage is connected to a second, round stage at the other end of the Garden by a rampway down the center of the floor—cutting through the general admission area on the floor. Above the ramp, running the length of the Garden is a giant metallic lattice structure, which turns out to be a huge two-sided screen 20 feet high and ~100 feet across.

The show kicks off with the opening track from the album which gets the crowd rocking. They pulled out some serious classics with "The Electric Co." (Friday), "Out of Control" (Saturday) and "Gloria" (Wednesday) each followed by "I Will Follow" ("Your eyes, make a *cir-cle...*").

"Vertigo" ("Hel-LO, Hel-LO!.... *HOLA!*") and "Elevation" ("EL-EV-A-TION!!!") had 20,000 people jumping up and down in unison.

The giant screen over the rampway then lights up for "Iris" (about Bono's mom) and shows a projection of Bono singing to and interacting with home movies of his mother. Then a stairway folds down and Bono climbs up into the giant screen which it turns out is a very long and narrow cage and we see actual-size Bono walking down "Cedarwood Road"—a song about his childhood street—as animation of his hometown scrolls past him. With a spotlight on him, you can see him through the video screen similar to what Roger Waters did with *The Wall* tour.

Later, with Bono singing on the second stage and the Edge playing guitar up inside the screen we see Giant-Video Bono spitting a stream of water onto actual-Edge (as seen below). Then holding Edge in the palm of his oversized projected hand. The screen is used for animation and projection of the band throughout the rest of the show.

For "Sunday Bloody Sunday," Larry straps on a single drum, and the whole band plays down on the rampway to pictures of the lives lost on Bloody Sunday up on the screen. The whole screen lowers to ramp level and shows a Johnny Cash video for intermission. The entire band sneaks into the screen and as the rarity "Invisible" starts up, you don't realize the band is back until the video screen starts opening up to show the four of them inside. For "Even Better Than The Real Thing" each band member is lit up inside the screen, superimposed with a giant video of themselves pixelating and breaking apart and reforming. Larry even has a full drum kit inside with him.

The Edge – projected across the arena-long screen. Captured by Lily Zampitella.

The band plays several songs down on the 2nd stage. And Bono pulls a girl out of the audience each night to dance with him to "Mysterious Ways." Two other nights they pull someone out of the audience to play guitar with them and then ***give*** them the guitar to keep. Again, different songs are rotated in. Friday night was the best version of "Bad" I've ever heard, with the crowd screaming along with Bono. "Isolation! – Desolation!... To *Let* it *Go!*... And so to find a wa-ayyyyyyy!... WIDE AWA-AAAAKKKE! I'M WI-IIIDE AWA-AAAAKKKKKKEEEEEEE!"

Saturday, we got "Angel of Harlem" ("This love won't let me *GO-OOO*! So LONNGGGGG. Angel of HAR-LEM!"), and Wednesday we got the gorgeous "Stuck in a Moment."

Even the songs I wasn't thrilled to hear when they initially started playing, still swept you up into them. "With or Without You" just kept building and building in intensity until everyone is screaming along with Bono – "With or Without You! With or With-*OUT* You!!! HO! – O! – O! – OHH!!! HO!!! – O! – O! – OHHHHHHHHHHHH!!!"

During "Pride"—a song I've seen live so many times—they take the song to a level I haven't seen before, yelling "Sing for peace in Baltimore! Sing for Peace in Charleston and Ferguson! Sing for Peace in Boston! After the terrible attack of April 15, 2013, this city showed it is *So Strong* and nothing will take your PRIDE!!! SING!!! IN THE NAMMMMMME OF LOVVVVVVVVVVE!" With "BOSTON STRONG" splashed across the screen in giant 20-foot-high letters. Absolutely amazing.

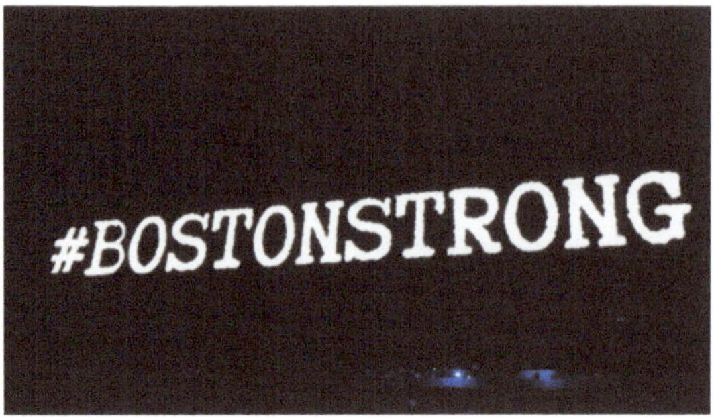

Bono singing for Peace. Pics again thanks to Lily Z!

I don't know how they do it, but I think U2 just keeps getting better and better with every tour. Twenty thousand people and it still feels like they're connecting with each and every one of us.

After "Bullet the Blue Sky" he talks about how much he loves the idea of America and how we've saved nine million lives in Africa with our humanitarian efforts for AIDS. And how in a few years if we continue, not only will we have reversed the spread of AIDS, but we will have eliminated passing it from mother to unborn child. For this, they sing Paul Simon's "Mother and Child Reunion" and thank us—"because if you're an American and pay taxes, then you're an activist." It's so rare to see someone thanking America for the good it does, instead of focusing on what it gets wrong.

The closing numbers and encores just brought the house down. "Beautiful Day"—forever entwined with the Patriots' 2002 Super Bowl victory. My favorite—"City of Blinding Lights" ("Oh – You – Look – So - BEAUUUUUUU-TI-FUL! – TONIII-IIIGHT!!!") with poles of white lights coming up from the ramp and down over stage and down from the screen. "Where the Streets Have No Name" again has everyone jumping up and down and "One" has all of us singing "*LOVE* is a temple! *LOVE* the HIGHER Law! *LOOOOOVVVVEEE* is a Temple!"

The final song is "40" with all of us singing… "HOOOOWWWW LONNNNGGGGGGGG! To Sing *THIS* SONNNGGGGGGG!"

Did I say Amazing? Amazing…

Thanks to Melissa for hooking us up!

Thanks to U2 for several of the best nights we've had—and that's saying a lot with my current concert count now at 358 concerts.

And thanks to any of you who bothered to read this all the way through. Hope you enjoyed it!

Until Next Time!

Story 5

So that all happened because of Melissa. Thanks, friend!

I promised that you'd all get introduced to her later. So I guess this is as good a time as any to do that. Here's a little window into how she came into my life—written at a time when she had just gone through major heart surgery.

I love that she was actually a part of our lives for weeks before she ever even met us. Also, I love that she later fully embraced a nickname we had for her, instead of taking it as a slight—which speaks to her great sense of humor and maturity, in contrast to our utter childishness at the time.

When your family doesn't live up to the unspoken promise of what that word is supposed to mean, your friends become your true family. This is just one expression of love towards just one member of that true family of mine. This one's for you, Melissa. You are a great person and a great friend.

Sweet Melissa (left) and Julie, way before my time with them began.

Age 15, Grade 10, 1982
"Flight of the Amazon Woman"

It was high school, so of course, anything that was different from the norm was subject to mockery. Walking the halls of Andover High, you could find the usual assortment of pubescent hormone machines. But if you were in the right place at the right time, you could see Melissa. Actually, she was so tall you couldn't miss her. I mean, she was *really* tall. In fact, she was so tall that her head and shoulders appeared to float on top of the throng of the crowded hallway.

Scott and I would see her pass and we'd call to each other in hushed tones, "There she is. Amazon Woman!" We were careful not to speak loud enough to be heard over the crowd for fear of those long arms that might lash out despite her apparently pleasant demeanor. We made a game of being the first to spot her. In fact, we soon envisioned her spreading those long limbs and taking to the air. After that, we were making bird imitations whenever the Amazon Woman flew by. "Caw! Caw!"

(I always took credit for coming up with the name "Amazon Woman." But it was my dad who first muttered, "My God, look at that *Amazon Woman*," as some gargantuan female strolled across his TV screen. Just thought I'd come clean here about that.)

So anyway, one day here we're hanging out at Donna's locker trying to be cool, and what happens? Donna says, "Oh, hey. There's my friend, Missy. Missy! Come here!"

Scott and I, of course, were looking at Donna at the time and were oblivious to Melissa's graceful approach. Next thing you know, there she was in full splendor hovering over us all in such close proximity. Scott and I locked eyes and, in a mixture of apprehension and surprise, simultaneously said to each other, "It's AMAZON WOMAN!!!"

Melissa soon became, and still is, one of the best friends I have. She just successfully came out of a heart valve surgery and is doing well. Keep flying, Amazon Woman. I love you.

Chapter 13

I originally wrote this next one right after it happened as one of my "stories." But it's about what happened next in the crowdfunding story. So, is it a **Story** or is it a **Chapter**? And hey, it's kind of a concert review too. So, is it another **Rocking Out**?

Well. It's all of the above, really. So, I'm titling it as all three. I'm mad with power. Try to stop me. You can't. Look ahead. It's already done. See… *Chapter. Story. Review.* Three for the price of one.

There's a bit of rehashing in this one. This is because it was a story that I sent out to everyone I knew in both my personal life and everyone I worked with at Polaroid. You see, a lot of these stories were written as "Easter Eggs" that I would send out to my entire building at work over 20 years ago.

As the first winner of Polaroid's "Safety Person of the Month" award (for instituting a Laser Safety program and training everyone in safe laser usage), I was awarded one of those silly little awards that didn't come with anything of value really. It did, however, allow the winner each month to park in a reserved parking spot right next to the front door. This was both a blessing and a curse though, because if your car wasn't in that spot when the bosses left, then they knew that you had left before they had. *D'oh!*

After my month of primo parking was up, I was nominated to be in charge of continuing the tradition and selecting and announcing the winner each month. This continued for a good several years after that. Each month I would send an e-mail to our entire building announcing the winner. What most people didn't realize was this… A lot of those announcement e-mails contained a screen worth of blank space (carriage returns) and then one of these stories that I added below that. At first, almost no one noticed. Then after some time, word got around and I had a lot of nice conversations about the things I had written.

The "Big Wheel" story was a hit with just about everyone that found it. Even people that I rarely talked to went out of their way to say they really enjoyed reading it. That made me happy. Well, this is another of those *Safety Stories*. But it also picks up right where we left off in our crowdfunding tale.

The Tour Fund was a success and the tour arrived in North America as promised. This story is about the first of three shows we saw on that historic tour and a special gathering beforehand for those of us that contributed.

Chapter 13 – Story 6 – Rocking Out #3
Age 30, 1997
"This Strange Tour" or "The Idea Man" (You Decide)

Let me tell you a story of a 19-year-old boy who discovered a band that he immediately fell in love with. The lyrics held vivid imagery, subtle wordplays and brilliant metaphors. The music was pure emotion with long instrumental passages and attention-grabbing time changes; songs that weaved together to tell a story.

Over the next 11 years, this boy grew to become a man (or so he'd like to think). He listened to a lot of other music but he always returned to Marillion and responded with his heart and soul to what he heard and felt.

In early 1997, I heard on the Internet that Marillion would not be able to tour the US again because of a lack of record company support. They had lost about $30,000 the last two times they came to the states.

Frustrated, I suggested that all the fans on the Marillion list server throw some money in a pot and create a "Marillion Tour Fund." Pledges poured in and a fund was established. Over $40,000 was collected in a few short months. The band was amazed. The record company agreed to sponsor a US tour in conjunction with our "Tour Fund."

In July the record company filed for bankruptcy, but the band still wanted to come and play. So the tour fund stepped up as the *sole* funding for the dates that the record company had booked but not paid for. As a thank you to the fans, the band promised to record a US concert and give 1,000 numbered, autographed copies of this CD to everyone who donated $25 or more to the fund. In addition, they would hold a special convention before their first show in Pittsburgh on Labor Day weekend. Special song selections and surprises were promised... naturally, I'm on a plane to Western Pennsylvania.

What a day it was. I was mesmerized seeing this band up close in the soundcheck. I was just to the right of center stage and they were playing these *great* old songs that I never expected to hear live; even taking requests from the audience. Could it get any better than this? I didn't think so, but I was wrong.

The band takes a break from the audience requests and presents a gift to Jeff Woods, the guy who set up an account for the tour fund, collected the money and dealt with the record company and later with the venues directly. Jeff W. receives much well-deserved applause.

Meanwhile, my buddy Julie calls over the lead singer "h" (Steve Hogarth). "This is Jeff Pelletier," she tells him. And I'm thinking, "Oh, he's not going to know who I am." But h's face lights up and he says, "The Idea Man!" He sticks out his hand.

"Pretty cool," I think and go to shake the hand. Next thing I know, I'm *yanked* up on stage!

"Get up here!" he says. H grabs the mic and introduces me. What a great feeling when I waved to the crowd and was greeted with enthusiastic applause from hundreds of people who love this music as much as I do. I must have looked like a blushing moron up there, but I didn't care.

H gives me a big bear hug (of which Julie took a great photo). Then Mark Kelly (the keyboardist) runs over to shake my hand, another great photo-op. Next, I was presented with a Thank You gift and before I even saw what it was, h said, "leave it with us and we'll all sign it." Cool!

On stage at The Graffiti Showcase in Pittsburgh with h (above) and Mark (below). Nice job being quick with the camera, Julie!

H said I didn't have to leave, so I plop myself down right there on the edge of the stage, between him and Steve Rothery (the guitarist). Julie climbs up and sits next to me and we're treated to a special acoustic set just for the fans, including several new songs and a few great Beatles covers. And there I was on stage with them. They had video cameras rolling so I may end up on a concert video somewhere.

The present turned out to be a nice list of names and *Thank You's* plus an envelope full of autographed pictures of each band member and a group shot.

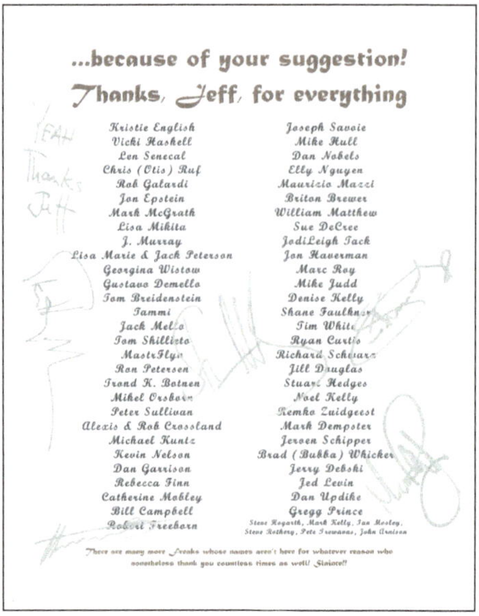

The band then spent a few hours chatting with the fans one-on-one—signing CDs, taking photos, eating, drinking and just having a great time. All that plus the full-blown concert that night was amazing.

And as if that isn't enough, The Chicago Tribune interviewed me for a feature article on Marillion, their unique tour and the power of the Internet.

Their new CD, *This Strange Engine*, is released today on their new US record label. So, run out and get it now. You won't be disappointed!

Chapter 14

More on That Night

That was what I wrote about that night 24 years ago. It's a pretty good re-telling of the main event. I can't help but add some extra exposition now though.

I can't remember if we were even supposed to be given access to their soundcheck that day. In the lobby, where the pre-show gathering was taking place, we had mingled and gotten to know a lot of Marillion fans in attendance during the day. Then at one point, we were standing near the entry doors when I heard this beautiful, beautiful sound coming from inside the main hall. It was h running through a piano/vocal version of "Cover My Eyes" and his voice is just *resonating* magnificently. We stood right there, *transfixed*, and listened to a distant sample of what we would be in for that night.

I had only seen Marillion live that one time before. That was two years beforehand, and in those two years, I had lived and breathed my obsession with this band. And here we were about to witness the start of a tour that I had helped launch. What a ride we were about to embark on...

We asked the security guy if we would be able to go in and he initially shook his head, No. But halfway through the next song, he ducked his head through the door, confirmed something with someone inside and then said, "OK. You guys can head on in." He opened the door and we all walked into the main club as the band was playing. At that point, we were treated to a full concert's worth of music—in the midst of which was the presentation I described in the story. The gift that was given to Jeff W. was a very nice leather jacket with the Marillion logo on the back if I remember correctly.

When h so unexpectedly pulled me up to join him, Julie remembers me saying (just loud enough for her to hear), "*Oh Jeez*!!!" She also told me how Mark made an absolute bee-line over to me to shake my hand and introduce himself. I know I must have stood up on that stage for a minute longer than I should have, but I was honestly flustered beyond all coherent thought at the time. Eventually, I regained my wits and took a seat with Julie on the lip of the stage as described in the story. Sorry to anyone out there who was thinking, *am I going to have to look at this jackass on stage for a whole song?*

If it was up to h, this bonus concert probably would have continued for hours.

Mark Kelly kept insisting that they stop in order to save h's voice for the show that night. H, however, just kept starting one song after another. Mark finally succeeded in reeling him in after well over an hour.

Signed pictures from the "Thank You" package I was given on stage—and then given again after they signed them.

After the show, Mark Kelly insisted on buying me a beer. In those pre-microbrew days, I didn't drink non-imported beer. Everything sudsy that was made in the USA pretty much sucked back then in my opinion. So when Mark asked me what I wanted, I said, "*How 'bout a Heineken*?" He asked the bartender and they didn't have it. Mark said he was sorry; they didn't have any. "*How 'bout a Beck's?*" I asked next. When that was another swing and a miss, Mark looked apologetic and said they didn't have that either. Realizing I was screwing up this nice gesture on his part, I quickly blurted out, "*OK. Anything imported!*" Here was a member of Marillion (my *favorite* band) trying to buy *me* a beer and I was making him feel bad about it. Ha. What a moron I was.

We had quite a few nice conversations with the band members that afternoon and evening. Ian Mosley (the drummer) was as laid back and friendly as could be. He even seemed genuinely pleased to hear me say that I really enjoyed the IRIS album (*Crossing the Desert*). We talked to h and I gave him a paperback copy of my favorite book that I had brought along with me: John Irving's "A Prayer For Owen Meany." I remember thinking they must have of a lot of time to kill on that tour bus, and maybe the book would pass through a few of their hands in their travels through our many states. H did say that *Steve Rothery really likes his stuff.*

[*Author's Note:* Coincidentally, I just listened to Chapter 85 of *The Corona Diaries* podcast, in which h's diary excerpt includes the name of a book that he's reading in Germany later in 1997. Looks like that paperback outlasted my expectations (kind of like the stuffed armadillo did in the book) and made it across the Atlantic and lived to fulfill its purpose.]

Two more pics from the "Thank You" package

At one point, Julie and I were eating at the same table with bassist Pete Trewavas. Julie took the opportunity to just *grill* him with questions. She asked him what his favorite track to play was from each album one after another and many other questions. Pete was a great sport about it and answered every one of those questions with generous detail. Each table around ours was like an audience for this impromptu interview session, eating and watching in silence. Three things I remember from that conversation were:

- Pete talking about how "Dry Land" is one of the hardest songs for h to sing (something I hadn't considered before). Whenever we saw him sing it in the future, I always took notice of the incredibly impressive vocal gymnastics he had to perform to get through that sucker.
- How the one song from the Fish era that Pete thought h could do a great job with was "Cinderella Search" (a song he eventually *did* sing). And Pete was absolutely right. He nailed it—even though they left off the "*Wel-come... Back to the Cir-cus!*" section at the end.
- Me asking Pete if they were going for a "Penny Lane"-type sound with the French horn in the song "Eighty Days" from the new album, and Pete excitedly saying that that indeed was *exactly* what they were going for.

Sadly, the night was brought to a quick end for the UK residents. We were talking to Mark again after the show when someone came over, pulled him aside and despondently said, *Diana's dead.* In no time, the band and crew all disappeared to confirm the heartbreaking news. Yes, this was in fact the night that Princess Diana was killed in an automobile accident. It was an abrupt ending to a night filled with great memories beforehand; and completely understandable.

Julie spent the late-night hours in our hotel room watching the news coverage as well. Meanwhile, I noticed that the film in my camera had jammed and wouldn't rewind itself. Thus, I spent close to an hour under a light-blocking tent of my bed linens, freeing up the film (millimeters at a time), eventually succeeding without breaking the strip and hand-cranking it back into its safe, opaque cartridge. (A task I thankfully completed without exposing and ruining the film, resulting in the two pictures of me on stage you saw a couple of pages ago, among others.)

Staring down from this high window, At the faces in the line
Cold from hours of waiting, How many people can you love?
When you're black and blue with bruises, From collisions on the road
The friction grind of traveling, This is the never ending show

All over the world in eighty days, Head in a blur of information
What kind of a man could live this way, I do what I can, I do okay
But right now, All I want to do is get real, if that's allright with you.
Right now, All I want to do, Somehow, Be myself with you

- Lyrics by Steve Hogarth & John Helmer

Chapter 15

Taking Notice

Julie and I saw three shows on the *Strange Engine* "Tour Fund" tour. We saw the Pittsburgh show (and pre-show convention) you just read about. We made the two-hour drive to New Haven, Connecticut, (Julie's previous town of employment) for their regular stop at Toad's Place. And we saw our local show at the Somerville Theatre (conveniently outside the city to the northwest of downtown Boston).

And would it surprise you to hear that we still paid less than $20 each for all three of those shows? Those tickets were $18, $11.50 (*that's right, **$11.50!***) and $18.50 respectively. How could anyone make a profit on an international tour at those ticket prices?

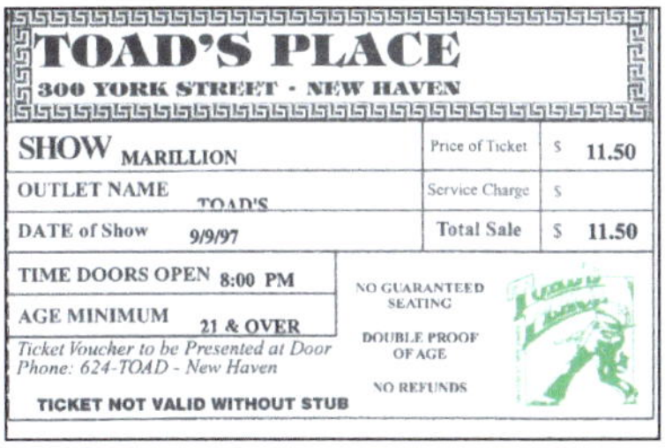

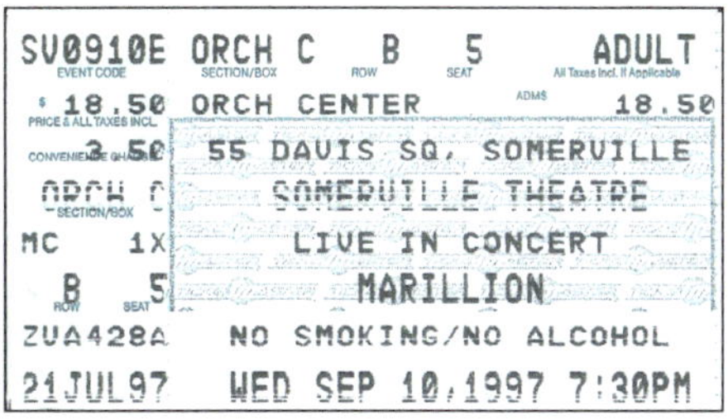

The Somerville Theatre has hosted some excellent bands over the years. I've seen Joe Jackson, Clannad, Sarah McLachlan, Porcupine Tree, Crowded House, Aimee Mann & Peter Wolf there, among others. It's also a functional movie theater. And would you believe on concert nights, they still show movies? ...in the same building! Insanity.

Hey, honey. Let's go see Good Will Hunting. Who cares if there's a full-blown rock concert happening right on the other side of the theater wall...

For our local show, I was the first one on the phone to the theater when tickets went on sale. I scored "pride of place" tickets for Julie and me in the front row center right under h's mic stand. Once that was taken care of, I asked what tickets were still available in a large block. I ended up buying eight additional seats a couple of rows behind us. I then proceeded to demand that all of my best friends see this concert. *No excuses!* This was going to be history in the making and they had to be there.

I distinctly remember gazing back at my friends several times throughout the show and they all seemed to be enjoying the hell out of themselves. I thought I had made some progress with them finally. But alas, not one of them ever bought a single Marillion CD to this day. While I didn't have any concrete success with my friends, it did seem like the world was starting to take notice though.

Julie's brother Mike was a news editor for the Chicago Tribune back in the '90's. One day, shortly before the tour had arrived on our shores, he was speaking to his little sister on the phone. When he heard that she was going to be flying to Pittsburgh just to see a band, he asked her the obvious question… *Why?* Julie brought him up to speed on the whole Tour Fund story and what I had started. Mike was so impressed he immediately brought the story to the Tribune the next day.

Mike gave my number to one of their staff writers and soon I was talking to a major American newspaper about my favorite band. I said they had to get in touch with Jeff Woods and get his perspective on receiving all of the donations. I also said he should talk to Mark Kelly as he was our conduit from the *Freaks* list server to the other band members.

The Arts section of the Chicago Tribune ran an article about this incredible thing that was happening with Marillion and their crazy fans. The article was written by Dean Golemis on September 23, 1997. The full feature can be found here:

http://tinyurl.com/tourfund

ROCK

British band's U.S. tour is computer-generated

By Dean Golemis
TRIBUNE STAFF WRITER

It was an outpouring more suited to large charity organizations or disaster-relief efforts, both in spirit and in financial success. Yet it was motivated by the simple, unwavering desire of a select group of cyberspace denizens determined to see their favorite rock band perform in this country.

How disappointed these fans were upon learning through the digital grapevine last February that Marillion, a little-known progressive rock band from Britain with a rabid cult following in the United States, would not be touring here in the summer. The gradual dissolution and eventual bankruptcy of the group's U.S. record label, Red Ant, had made it all but impossible for them to cross the Atlantic.

Immediately, the Marillionites took to their computer keyboards. Messages were posted on the Internet announcing a fundraising effort, bulk e-mail soliciting donations was sent, and by summer the British quintet had finally found its ticket to the U.S. — paid for in full by donors worldwide using the Internet as the sole means of raising cash.

Thanks to these dedicated fans and the $47,000 they have garnered so far — far ahead of the original $30,000 goal — the band put

SEE MARILLION, PAGE 3

Marillion

CONTINUED FROM PAGE 1

together a 21-date tour of the U.S. in support of its ninth studio album, "This Strange Engine." The tour wraps up Tuesday night at the Park West.

Although music fans have always backed their favorite performers in various ways, such a massive outpouring by Marillion's faithful through the boundless outreach of the Internet is unprecedented.

Marillion's on-line followers, known as Freaks (they take their name from the lyrics of one of the band's songs), tout themselves as computer-savvy music fanatics who discuss the group on the Internet's World Wide Web (www.marillion.com) and Newsgroup message boards (alt.bands.marillion). It was at these electronic watering holes where Jeff Pelletier, an optical engineer in Bedford, Mass., conceived the fundraiser to bring Marillion to the States.

Pelletier credits the "instant communication" of e-mail and on-line forums with the fund's success. When he proposed his plan to his fellow Freaks, the response was overwhelming. "People were giving money just to see this happen, because it was such an amazing thing," he recalls.

The band members, on the other hand, were skeptical. Though not averse to the idea, they thought that the enthusiasm would soon cool, and that they would be perceived by the media as beggars. However, it wasn't long before the crusade was drawing modest contributions from fans all over the world, including South America and Australia. One anonymous British donor even forked over $1,300.

"We were taken up by the fans' enthusiasm," says Mark Kelly, Marillion's keyboardist and a frequent contributor to the Freaks' message board. "It just kept going and it didn't fizzle out. Once it got past the $25,000 mark, we started making plans for the tour."

Kelly and his bandmates — Steve Hogarth (vocals), Steve Rothery (guitar), Pete Trewavas (bass) and Ian Mosley (drums) — have maintained a cult following in this country since their 1983 debut album, "Script for a Jester's Tear," which then featured flamboyant singer Fish on lead vocals

'We were taken up by the fans' enthusiasm. It just kept going and it didn't fizzle out.'
Keyboardist Mark Kelly

and Mick Pointer on drums. Outside their tight fan base, they have been mistakenly labeled here as a heavy metal outfit. In fact, Marillion weaves a tapestry of sensuous melodies, fluid dynamics and shifting time signatures.

After the fund was established, Kelly handpicked a friend, Jeff Woods, to organize the fundraising and collect all the money. (This direct involvement by a member of the band was a convincing gesture to the Freaks that the fundraising drive was not some fraudulent money-making scheme spreading through cyberspace.) Woods, a software developer and Internet service provider in Morrisville, N.C., then took it upon himself to spread the word and print posters and fliers heralding Marillion's return to the States. He also handled promotions for the band's recent shows in Rochester and Poughkeepsie, N.Y.

"I was nominated [to head the fund] because of my association with the band," says Woods. "I am known on the Internet as a fan who knows the band and who goes to a lot of shows." Since 1992, he has traveled to see Marillion 46 times and even proposed to his wife on the group's concert stage in London last year.

Though it has enjoyed success in Europe, Marillion has suffered financial trouble on previous U.S. tours. According to Kelly, the band lost $30,000 during its last visit here despite partial backing from its record company at the time, El Dorado. That tour, in 1995, visited only the East Coast and the Midwest. West Coast fans who had not seen the band since 1992 were treated to three shows in California on the current tour.

Also falling victim to Red Ant's financial woes was "This Strange Engine." The album was originally slated for record stores earlier this summer but has been available here only as a special-order import. It will finally be released on the small Eagle Rock/Velvel label on Oct. 7.

As a thank-you gesture to all of the fund's contributors, Marillion will send out up to 1,000 autographed copies of a live CD recorded during one of the U.S. shows and paid for by money left over from the tour fund.

"Our fans are an honorable bunch," Kelly says.

The article was picked up by the AP wire and appeared in several papers around North America (including the Buffalo News on September 26, 1997). This was no small thing for Marillion fans on this side of the Atlantic. The band usually flew completely under our radar, but now they were getting some serious word-of-mouth promotion. A copy of the article was sent to me by a local Buffalo fan and appears on the next page.

26 GUSTO THE BUFFALO NEWS/FRIDAY, SEPTEMBER 26, 1997

POP MUSIC

Net power

FREAKS TO THE RESCUE

U.S. fans save the sounds of Marillion

By DEAN GOLEMIS
Knight-Ridder

IT WAS AN outpouring more suited to large charity organizations or disaster-relief efforts, both in spirit and in financial success. Yet it was motivated by the simple, unwavering desire of a select group of cyberspace denizens determined to see their favorite rock band perform in this country.

How disappointed these fans were upon learning through the digital grapevine last February that Marillion, a little-known progressive rock band from Britain with a rabid cult following in the United States, would not be touring here in the summer. The gradual dissolution and eventual bankruptcy of the group's U.S. record label, Red Ant, had made it all but impossible for them to cross the Atlantic.

Immediately, the Marillionites took to their computer keyboards. Messages were posted on the Internet announcing a fund-raising effort, bulk e-mail soliciting donations was sent, and by summer the British quintet had finally found its ticket to the United States — paid for in full by donors worldwide using the Internet as the sole means of raising cash.

Thanks to these dedicated fans and the $47,000 they have garnered so far — far ahead of the original $30,000 goal — the band put together a 21-date tour of the United States in support of its ninth studio album, "This Strange Engine."

Though music fans have always backed their favorite performers in various ways, such a massive outpouring by Marillion's faithful through the boundless outreach of the Internet is unprecedented.

Marillion's online followers, known as Freaks (they take their name from the lyrics of one of the band's songs), tout themselves as computer-savvy music fanatics who discuss the group on the Internet's World Wide Web (**www.marillion.com**) and Newsgroup message boards (**alt.bands.marillion**). It was at these electronic watering holes where Jeff Pelletier, an optical engineer in Bedford, Mass., conceived the fund-raiser to bring Marillion to the States.

Pelletier credits the "instant communication" of e-mail and online forums with the fund's success. When he proposed his plan to his fellow Freaks, the response was overwhelming. "People were giving money just to see this happen, because it was such an amazing thing," he recalls.

The band members, on the other hand, were skeptical. Though not averse to the idea, they thought that the enthusiasm would soon cool, and that they would be perceived by the media as beggars. However, it wasn't long before the crusade was drawing modest contributions from fans all over the world, including South America and Australia. One anonymous British donor even forked over $1,300.

"We were taken up by the fans' enthusiasm," says Mark Kelly, Marillion's keyboardist and a frequent contributor to the Freaks' message board. "It just kept going and it didn't fizzle out. Once it got past the $25,000 mark, we started making plans for the tour."

Kelly and his bandmates — Steve Hogarth (vocals), Steve Rothery (guitar), Pete Trewavas (bass) and Ian Mosley (drums) — have maintained a cult following in this country since their 1983 debut album, "Script for a Jester's Tear," which then featured flamboyant singer Fish on lead vocals and Mick Pointer on drums. Outside their tight fan base, they have been mistakenly labeled here as a heavy metal outfit. In fact, Marillion weaves a tapestry of sensuous melodies, fluid dynamics and shifting time signatures.

After the fund was established, Kelly handpicked a friend, Jeff Woods, to organize the fund-raising and collect all the money. (This direct involvement by a member of the band was a convincing gesture to the Freaks that the fund-raising drive was not some fraudulent money-making scheme spreading through cyberspace.) Woods, a software developer and Internet service provider in Morrisville, N.C., then took it upon himself to spread the word and print posters and fliers heralding Marillion's return to the States. He also handled promotions for the band's recent shows in Rochester and Poughkeepsie.

"I was nominated (to head the fund) because of my association

These articles happened thanks to Mike Hanlon (and of course, thanks to Julie Hanlon). Thanks, Mike! You're part of the story now.

And Dean Golemis himself (the author of the article) is also a Marillion fan as it turns out. Here's what he has to say *today* about learning *all those years ago* (as George Harrison would say) of what we crazy Marillion fans were up to…

A Note From Dean Golemis

"Back then we were just discovering the vast potential of the internet as a tool for reaching many people from everywhere. When I was asked to write the Chicago Tribune article about Marillion fans raising money to bring the band to the States, I thought, 'Oh, you can also do that on-line—how cool!'

"Being a Marillion fan myself, I was aware that the band had a small yet devoted following, as prog rock groups typically do. So, I was heartened to hear about the prospect of a U.S. tour that year.

"I am pleased, though not surprised, to hear that your initial outreach efforts have blossomed into a widespread business model. Of course, 'crowdfunding' is a household practice now. What you and fellow fans did back then was certainly groundbreaking, so many kudos to you!"

Dean Golemis
December 2020

Chapter 16

MARILLIONROCHESTER

The tour was a success and Marillion were true to their word. They said every donor would receive a CD of one full show from the tour and that they did. The set in Rochester, NY was selected and the CDs were printed with limited edition numbering on their inner ring. At the time, I thought The *Tour Fund CD* would be the perfect name for this unique prize. However, they chose to call it MARILLIONROCHESTER—all one word, all caps.

Mark Kelly then had the unenviable task of breaking the news to his bandmates that he had promised that they would all sign the first thousand CDs. Thus, all five members had to sit down with 1,000 of these gifts and sign each and every one of them. They completed the task (with cramping fingers I'm sure) and made a lot of their most loyal fans very happy.

Erik Nielsen (all-around Marillion tech and who also ran their shop and website) recalls the CD as follows when asked about it by Prog magazine:

"I remember the band promising a live CD to fans. This was just before the tour to help spur on extra donations, but we definitely sent it out some time after those dates ended as something we could give back to anyone who had helped."

As discussed, the Marillion Tour Fund of 1997 was the first successful internet-based crowdfunding campaign. And now the *MARILLIONROCHESTER* CD was the first physical product of an internet crowdfunding campaign.

A few interesting things to note about the manufacture of this CD. The printers that manufactured the CDs forgot to reset the numbering on their presses. So, there is no *#1 of 1,000*. The numbers started somewhere well over 2,000 and went up from there. This was too bad because even though people had contributed before I knew it was actually happening, everyone on the *Freaks* list had generously agreed that I would be given #1 of 1,000. I was disappointed about the numbering snafu at first. But honestly, it was a non-issue.

In addition to that mistake, they also spelled Marillion's name wrong on the CD itself (dropping the second "I" in Marillion). Obviously, the band couldn't have their name misspelled on an official release of theirs. Yet another CD pressing would definitely cost too much.

Tour Fund CD (left) – even though you can't make out the misprinted artwork in 2D. Mis-numbering shown at the bottom of the inner ring on the back (right). I got #2362. Anyone get a lower number than that?

So, the band agreed to have them just re-print the updated CD artwork on top of the original CDs (over the incorrect artwork). If you have one of these CDs and hold it up to the light, you can see the impression of the original lettering. Sure enough (if you look closely), you can verify that there's only one L in the title, MARILIONROCHESTER in the layer beneath the surface.

The liner notes for this exclusive prize start off by stating that "If you've got a copy of this CD, then you are part of the reason it exists." They give a brief history of how it came to be, how it was their first official release of an entire concert from start to finish without edits and expressed their gratitude and appreciation. Near the end they thanked a number of people including Jeff and Michelle Woods (for all their efforts), Paul Baines ("for the largest single donation"—*good one, Paul!*) and me, the person "who had the brainwave in the first place."

The band has this to say about this now Out of Print CD in their discography:

Rochester - Live Album: Released May 1998, Recorded August 1997

In the summer of 1997 Marillion played a four-week tour of North America. But what made this tour unique was that for the first time ever, a band's fans directly financed a tour. It all began when Mark [Kelly] posted to the internet in January 1997 that a North American tour in support of This Strange Engine was not likely, mainly due to financial reasons. By 1 March 1997, at fan Jeff Pelletier's suggestion, $10,000 of tour support money had already been raised. By the time the tour started in August, $40,000 was collected, and the fund was still looking forward; together, our fans from all over the world raised over $61,000 for the North American Strange Engine Tour.

As a way of returning the favour to all those who shelled out and put their money where their mouths (and hearts) are, we've released an entire concert from start to finish without edits. These CDs were sent to everyone who donated to and helped with the fund. Less than 2000 were pressed, and no more will ever be made. For those who did donate, tell your friends you made this record! We've always had a great vibe at the shows in America, Canada, and Mexico, but 1997 will stand out as a tour that really did belong to all of us - band, crew, and fans. In truth, everyone involved with the fund played as much of a part in it as we did. Thank you.

H, Steve, Mark, Pete, Ian

When my package arrived in the mail, I distinctly remember never being more proud or more appreciative of anything before in my life. Holding that CD in my hands felt so surreal. I was actually thanked *by name* by my *favorite* band on one of their CDs—one that only existed in the first place because of a single flap of my *butterfly* wing whose effect had cascaded around the world to result in this whole odyssey taking place.

At this point, I should note that another error occurred on this CD. Not only was *Marillion* not spelled correctly, but my first name was spelled with a G instead of a J (Geoff instead of Jeff). It was almost comical how bad Mark Kelly appeared to feel about it the next time we spoke. He said to me, "I'm so sorry we spelled your name wrong on the Tour Fund CD." I quickly assured him not to worry about it at all. And I told him how I had joked with anyone I showed it to that "Geoff with a G" was now my stage name. Ha.

---

*Over the following years, on the occasional blue moon, I would see one of these Tour Fund CDs pop up on eBay. I would always watch with interest as it would usually fetch a couple hundred dollars. Once I saw one pull in over 300 English Pounds. If you think about it though... They signed the first 1,000 CDs. But they printed fewer than 2,000 of them in total. So wouldn't it stand to reason that the CDs that **weren't** autographed are in fact **rarer** than the signed ones, and should be worth more? Counterintuitive, I agree. But logical nonetheless...*

The reason I will always cherish this CD is because of what it stands for. A group of like-minded people having the nerve to put their money where their mouths were—who had the trust to send their money to someone they hadn't met before—who were able to put all their petty differences aside for one beautiful moment in time and do something both selfish and selfless in the same act.

... And also because of a band that had the nerve to go against convention and listen to its fans and let them make a difference, (despite the fact that the misinformed or embittered could have easily decided to paint the Tour Fund story in a negative way for Marillion). ...A band that couldn't accept such a large amount of money without giving something back—and that continues to inspire me to this day with their creativity and their beauty.

Funny how one small action can change your life. I have now been lucky enough to get to know the members of my favorite band. And when I go to concerts here and abroad, there are people to this day who recognize me. I usually don't bring it up anymore when I meet new fans, because I don't want to seem like I'm expecting a certain reaction. Yet it's still nice when I meet someone new at a Marillion concert and they recognize my name and thank me.

Story 7

Let's take another breather and transition from stories of a man's favorite band to stories of man's best friend. This one has a bit of a different tone than our previous tales.

It is very true that the Pelletiers were in fact cursed with many *very* stupid dogs. It's worth mentioning though, that when we still lived with my mom, we had a german shepherd named Brutus. Brutus was a member of Dog Mensa. He wasn't stuck up about it or anything, but he knew his shit. He could learn anything you wanted to teach or train him to do. He could have made a good living tutoring other dogs, but he was content just patrolling his domain and lavishing affection on its residents. He wasn't lazy. He just didn't need the money. He didn't have any bills to pay and whenever he was hungry we fed him. So why work?

Brutus did have one episode though that he wouldn't look back on as one of his cleverest moments. He gave in to his base temptations one night after dinner. With two strong paws up on the kitchen table, he dug into a chicken carcass with abandon. Unfortunately, those crushed and splintered chicken bones were not as pleasant coming out as they were going down. Poor Brutus spent an entire night howling in pain on his haunches in the backyard as he shredded his poor ass passing those chicken bone splinters.

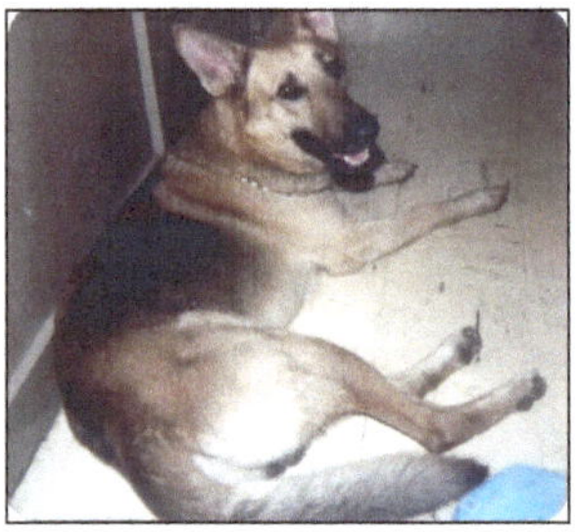

Brutus (left). Leroy (right). And (since we're talking about dogs in a book about Marillion), a very young h with his infamous bulldog (center). "He wore a red coat and walked a bulldog, saw them reflected in the mirror of the lakes."

Can't we help him?, I had asked at the time. Sadly the answer was a resigned *No*. There was nothing we could do but let him get through it. I've enjoyed the many pets we've had over the years. I've loved both our cats and our dogs. There was indeed something special about Brutus. This is not his story though. This one is about bad, bad, Leroy Brown. The dumbest dog in the whole damn town (of Andover).

Age 24, Grade 0 (0 for unemployed), 1991
"Leroy Brown"

The Pelletier family has always been cursed with stupid dogs. I don't know why this is... it's just something we've dealt with since I was in seventh grade when my older brother Alan and I moved in with our dad. The stupidest of the batch was "Beauregard" who couldn't let a second go by without barking. Yap, Yap, Yap... never-ending... Wouldn't shut up for anything.

My father even gave him one of those electric shock collars that zapped him every time he barked, but that didn't stop him either. He'd just sit out on his chain repeating the same cycle over and over again.

"Yelp!" ZZzzzzzt. Pause...

"Yelp!" ZZzzzzzt. Pause...

"Yelp!" ZZzzzzzt. Pause... "Yelp!"

He never quite put two and two together.

But anyway, "Leroy Brown" was the latest model in a long line of Pelletier dogs that never were able to outlive the family cat. Leroy and I had gotten off to a rocky start a year before when he first arrived at our fine home. I wouldn't allow him into my bedroom until he was properly trained. You know why. Also, I was developing an allergy to something and I suspected it to be him.

One day, I was lying in bed reading with the door closed, when Leroy discovered that the door may have looked like it was closed, but it hadn't actually clicked shut.

He poked his head into the room and wagged his tongue at me.

"NO!", I commanded from the bed with my most authoritative voice. But Leroy must've been harboring some serious resentment against me and my impenetrable fortress. He seized the opportunity while it presented itself.

He sprinted forward into the room, bolting straight toward me at an alarming speed. At the last possible moment (with me rolling over and ducking for cover) he spun around tracing a tighter circle than one would think possible at that velocity. And as he did this, he spewed forth a spray of liquid dog crap that painted the side of my bed and the rest of my room across a 180° arc. Then he tore out of the room before I could begin to comprehend what had just happened.

But that was then and this is now and he turned out to be a pretty damn good dog, though a little slow at times. So I went down to the garage to let him in on a cold rainy day.

He was just sitting out in the middle of the back driveway with his head on his paws (legs splayed out like a platypus) letting the rain pelt all over him. I tapped on the glass pane in the door and Leroy's head jerked around to look at me. Of all the dogs we've had, Leroy was the most anthropomorphic. You could plainly read his expressions and what they meant. He looked up at me as if to say, "You're gonna let me in? ALL RIGHT!"

He got up and bolted in through his "doggie door." He was jumping up and down at the end of the chain in a virtual frenzy. "All right! Let me in! Let me in! Man, it gets lonely out there and all that water keeps coming down out of the sky and I don't know what the deal with *that* is, but I don't like it!"

I stood just out of his reach not wanting him to jump up and get wet dog slime all over me. But then I said the one thing to get him even more excited... "Want to come in?"

"What the heck do you think I've been *saying* here?!?," Leroy replied and commenced to jump about even more frantically. "Of course I want in! Let's go! Let's *go*!"

At this point, he backed up a step and started to do the doggy-shake. It happened so fast I didn't have time to react. It started with the head (ears-a-flappin') and traversed down the spine to the tail which was already wagging at top speed, showering me with dog-scented water.

My fault, I thought. I should've seen that coming a mile away. *The dog's wet, so he's gonna shake it off, you idiot.* I tried not to get pissed about it. I had that "wet dog" smell all over me now and would probably be sneezing uncontrollably soon, but it wasn't Leroy's fault. I should have known. I reached down to unhook the chain from his collar and he slobbered all over me. "Love you, love you, love you!" he conveyed with each lick.

I set him free and he darted to the door leading from the garage into the basement. "Come on, come on! What's taking you so long, Jeff? Jeez, I've been waiting here forever! Come on! Open the door! You know I can't open it, I don't have any *thumbs*. Come on, hurry *up*!" He bounced up and down in front of the door looking at me, then the door, at me, then the door, me, the door... getting his message across loud and clear.

"OK, OK, I'm coming," I said. I reached for the doorknob and was sure that the dog was going to smash his head into the door as I pulled it open—thus closing it again, preventing the entrance he'd been so excited about. Though the dog had learned a couple of things at least and pulled back just enough to let the door open and then bolted up the stairs.

Leroy was fun to have around when you get right down to it, even though he was definitely the dunce of the litter. He was a cute friendly mongrel that didn't bark too much. However, his bladder still hadn't learned to talk to his feeble brain and it took him quite some time to figure some things out.

One of those things was the difference between the cat's food and his food. Leroy hadn't yet learned that eating from the cat's bowl was forbidden. So up the stairs he runs, and what does he do? He dives straight into the cat food. Gulp, gulp, lick, smack, gulp, lick...

"Leroy, NO!" I yell and grab a newspaper to whack him on the nose–a punishment that doesn't really hurt but he hates it anyway. He sees the rolled-up paper and flees for his life. Around the kitchen island I chase him and then into the living room. Around the living room table and down the hall to the bedrooms. At the end of the hall, all the bedroom doors are closed and he makes an about-face.

"A-HA!" I proclaim with triumph in my voice. "Got you *now*!"

To which Leroy responds by letting loose with his bladder. Doggy-urine comes streaming down like he hadn't gone in days. I throw away the paper and yell, "No, Leroy! STOP!!!" But out it keeps coming, with him just looking at me saying, "Sorry. Can't stop once you've started. You know *that*."

After what seemed like three gallons, he stopped. Such is life with a Pelletier dog.

Story 8

I have Julie to thank for this experience. (Who are we kidding? When do I *not* have Julie to thank?)

Julie has been a Sting fan for as long as I've known her. While she was still living in Branford, Connecticut, and working at Yale, she called me one day at work with a question. I have a chance to buy tickets through his fan club to a special Sting show in New York City, she said. It's a small club, and he's going to play the whole new album. If you drive down we can take the train into the city. What do you say?

I wasn't sure I wanted to go all the way to New York. These were the days before I started traveling for shows. But she said she really wanted to go and wasn't going to do it alone. So after only a small amount of twisting of my upper limb, I agreed. Good decision, Jeff. Great suggestion, Julie.

Re-reading this story all these years later, I chuckle a bit embarrassedly at my star-struck gushing. But hey, life is for living and enjoying. And it's pretty clear I was enjoying myself fully back then.

You should probably know beforehand that I was (and still am) a complete Peter Gabriel fanatic. I have often said that my favorite non-Marillion concerts of all time are a tie between Roger Waters' *Radio KAOS* tour in 1987 and Peter Gabriel's *Secret World* tour in 1993 (both making local stops at the Worcester Centrum). His *SO* album came out in the spring of '86, near the end of my freshman year in college. Throughout my sophomore and junior years at U. Lowell, he dominated the airwaves of New England in a wonderful joining of popular and progressive music.

My roommate, Mark Gillette ("the best a man can get"), gave me a five-foot-tall poster of that album cover as an unexpected gift. Senior year, when we moved out of the dorms and into the apartment over the arcade on University Avenue, that poster got *pride of place* in our kitchen, facing the big windows that looked out over the north campus. On more than one occasion when someone realized where I lived, they would say, "You mean the apartment with the huge picture of Peter Gabriel? *Cool.*"

With that level of devotion on my part now established, you may feel free to send some good-natured and well-deserved mocking my way if you so choose. I'm always up for a good joke at my expense.

Age 29, March 1996
"Sting and Pete"

Music has always seemed to mean so much more to me than those around me. So some of you may not care much to hear about this, but I'm going to write about it anyway. To *me*, it was unbelievable ...

There's a little theater in Brooklyn called The Academy where they recorded a lot of the MTV Unplugged concerts. And this Saturday night, I was there to see Sting in concert. This is a man who easily packed the largest stadiums when he fronted "The Police" and still sells out the large venues in his solo career. So this is no trivial thing to see such a well-known and celebrated musician in such an intimate setting.

We were all there at this special show to be the first audience to hear all the music from his new album, *Mercury Falling* (available later this month). The concert was phenomenal, but that's not all there is to tell.

Before the show, Julie and I were ranting to one another about how we couldn't believe that we were standing less than 10 feet from the stage when a murmur floats across the floor. People started pointing at the reserved balcony seats (up behind the floor). I turn around and a familiar profile is taking his seat accompanied by security. I'm trying to place the face and then I get a good look and I catch my breath.

Who is in the same room with me less than fifty feet away but *PETER GABRIEL*!!! Everyone on the floor is pointing up and saying, "Pete, Pete," *murmur, murmur,* "Pete, Pete." I couldn't help myself; I raised my hand and yelled over the crowd, "PETER!" He turned and looked right down at me, then smiled and waved. Then he half-stood (almost embarrassed by the attention) and waved to the rest of the crowd. Everyone went nuts.

I must admit, I almost wet my pants. Peter Gabriel has been making progressive, intelligent, ground-breaking music for over 25 years now—first as the front-man for Genesis and now as a legendary solo artist. So to have made this contact with him was the coolest thing that's happened to me in a while. He's definitively my single favorite and most respected solo artist.

This was the high point for me, but that wasn't all that happened. As soon as we calmed down about Peter being at the concert, Robin Williams walks in. The crowd goes wild again. Robin runs back and forth, does a few antics in that manic manner of his, bows to more applause and sits down near Peter. After him, Julie recognized Miles Copeland—Stewart's brother and manager of The Police among other bands.

The show starts and I can't believe how close Sting is and how good he sounds. A club show is so much cooler than an arena show. Then halfway through the first song, a door opens on a reserved balcony box, and who walks in but Brad Pitt. You should have heard all the girls on the floor, "Oh my God, that's BRAD PITT!!!" He was trying to hide under his baseball cap behind his girlfriend (and co-star from "Seven") Gwyneth, but once he realized that was pointless, he leaned forward and got into the show. Later on, I even saw him playing imaginary keyboards on the rail in front of him.

At one point in the show, Sting walked over to play right in front of us and, as the crowd surged forward, we were less than five feet away. My friend, Julie, was swooning and stunned speechless. She's been obsessively in love with the guy since the early '80's and there he was, within spitting distance.

So that's a tale of an exciting weekend. Call me a fool for being so star-struck, but hey, I saw a Sting concert in the same room with Brad Pitt, Robin Williams, and *Peter Gabriel*. My buddy Pete!

That Wraps It Up, Right?

All right. The Tour Fund was created. It was a success. The tour happened. I had a little moment of appreciation. And then the Tour Fund CD was printed and sent out to all the donors. Mission accomplished. So that's the end of the story, right?

Well, we haven't gotten to that stage in North Carolina yet, have we? And you've probably noticed that you're less than halfway through this book you're holding. (Or the % indicator at the bottom of your tablet is only around 32%). So, you would be correct in guessing that I'm not done blathering on. Maybe I'll just talk to you about my dreams for a hundred pages.

There are actually a few more things that happened beyond this point in our story that you might be interested to hear about. Several of them are personal to me. But one of them is a very significant landmark in the story of crowdfunding itself that we haven't covered yet. Let's start with one of those things that was personal to me. I mean, I'm the one writing this. So, it all has to be personal to me, right?

Chapter 18

Fan of Honor

How many of you know someone who's won an MTV video music award? Would you believe it if I told you I have? OK. Technically it wasn't an *actual* MTV award like "Best Video" or "Best Song." But I shit you not when I tell you I won an award at the MTV Europe video music awards ceremony at the Fila Forum in Milan, Italy in 1998. It was the *Fan of Honor* award given to me by Carlsberg beer, who had held an online "Best Music Fan" contest as part of their sponsorship of the video awards that year.

So OK. You're right. It's not the same thing as a genuine MTV Video Award. I never got one of those shiny astronaut statues. But I'm counting it anyway; thus making me 20% of the way towards achieving my "**MEGOT**." Which begs the question… I suppose it's possible I could win a Grammy for the audiobook reading of this text. But who's going to help me with the interpretive dance version of these stories to get me that sweet Tony and our caricatures hung on the wall at Sardi's?

This next story was originally published in the Web Germany Marillion fan magazine, *Made Again*. It was written as if speaking to people already in the Marillion-know. But you're up to speed now, aren't you? You're pretty much *in the know* now as well. I've made a few present-day edits for accuracy, grammatical correctness and additional flourishes of exposition. Other than that, here for your reading pleasure (I hope) is the *Fan of Honor* chapter. It's a bit of a wild ride. Strap in …

Once again, it all started with a post on the *Freaks* list—this time not from *me*, but from Marillion manager Lucy Jordache. She told us that Carlsberg Beer was sponsoring the 1998 MTV Europe Music

Awards in Milan and they were holding a "Best Music Fan" contest. Write a brief description of your favorite band and why you deserve to win the Best Fan contest and you could win a trip to the MTV Europe Music Awards show.

To paraphrase Lucy, there are a lot of diehard Marillion fans on this list. Surely someone out there has a Marillion story they could share, she prompted... *Well, why not give it a shot?* I thought (rhymingly). I had nothing to lose.

I went to the Carlsberg website, found the contest page and submitted an entry. I wrote about being in love with Marillion's music for over 10 years. I wrote about planting the small seed that blossomed into the Tour Fund. I wrote about attending This Strange Convention in Pittsburgh and meeting the band. I wrote about h pulling me up on stage and people actually applauding while I stood there feeling overwhelmed and overcome with appreciation for my fellow fans and for this great band. I wrote it all down and sent it in. And I said to myself, *I'll never win, but if I was judging this contest, I'd pick that one as the winner.*

Carlsberg, as you may or may not know, is based in Denmark. Amidst all the fine print, it apparently stated that this contest was open only to European residents. I didn't realize this when I submitted my entry. However, one day (after enough time had passed that I had completely forgotten about this contest), my work phone rang. Fortunately, I happened to be sitting at my desk to answer. In those days, this wasn't often the case as I was usually in the laser lab. So, this was fortuitous timing.

The person on the other end of the phone asked if this was Jeff Pelletier. I confirmed it was. He introduced himself as someone from the Carlsberg company and I immediately screamed in my head, *Holy Shit!* The judging committee for the contest had read my story and were blown away. They told me about the rules of the contest that was only supposed to consider people from the European continent. But they quickly said that it was such a good story that they wanted to make me one of the contestants regardless. It was a story that needed to be told.

Marillion was contacted. The Tour Fund story was verified. A short time later I was told that I was heading to Milan.

Three European entrants were chosen as the finalists for the Best Fan contest. In addition, they decided to create a special category just for me—"The Fan of Honor." I wouldn't be part of the actual voting contest because of the contest rules, but I would take part in all the activities under the newly created title. We were all going to make the trip to the MTV Europe Music Awards in Milan—the three "Best Music Fan" finalists and the Fan of Honor.

Timing is everything. Believe it or not, I had already been planning a trip to England for Julie and me to see Marillion only days before the Awards Show was scheduled to take place in Milan. After learning of this, Carlsberg generously agreed to give me the price of the ticket to Milan in place of the actual ticket. This allowed me to put that money towards a three-legged journey across the Atlantic to England and then on to Italy before returning home to the US.

Since Julie was already going to be in England with me, and it wouldn't cost Carlsberg anything extra to have her stay in the same hotel room with me, they agreed to let me bring Julie along for the ride to the awards show at the Fila Forum.

(Additionally, the direct flight they were going to book for me cost as much as both of Julie's and my discount fares combined. So, our airfare and our hotel in Milan were completely covered. Thanks, Carlsberg!)

Julie and I were on our way to see our favorite band both in London and in their hometown of Aylesbury, and then on to Italy to expose MTV (and hopefully the musical world) to the wonderful secret joy of Marillion on an international broadcast.

What can one say about traveling abroad to see a Marillion concert that hasn't already been said before? The Aylesbury show was a blast and we actually met more Americans in the local pub before the show than locals. How is that even possible? Because we're all *Freaks*, that's how.

Many Marillion family members were in attendance since it was their hometown show. We met h's kids, Pete's wife and kids (and even his mother) as well as Mark's girlfriend at the time, Angie. All of them were warm, welcoming and completely open to chatting with the fans. Pete's mum was a hoot and one of the nicest people we talked to that day. When she learned how far we had traveled, she looked at us curiously and asked, "You came all the way from America to see my Pete?"

The London show was a whole 'nother experience. The venue was much larger and packed with way more fans than I was used to seeing at the smaller Marillion club shows in the States. A full light show was on display and the house was rocking with an absolutely killer concert. Needless to say, it was another amazing night of fantastic music amongst fantastic people.

And then the next morning we were off to Milan where we were once again reminded of the kindness of Marillion strangers. I had queried on the *Freaks* list if there were any local fans in Milan who wanted to get together. That first night in Milan we met up with Max and Marie Luisa who took us out for a great night on the town with some of the finest Italian cuisine I've had to date. We had such a great time with such easy conversation, that we all decided to get together again the night after the awards show. (Thanks, guys. Your kindness and hospitality are still appreciated to this day. And if you ever find yourselves in New England, we'll endeavor to return the favor.)

* *

The MTV Awards Show itself was certainly an experience. The music—well, *c'mon* ... it's MTV. What did you expect? What is currently hailed as the best in popular music paints a very sad picture of the music industry today. But R.E.M. was a rare breath of fresh air amongst their "peers" that night, putting forth a stirring version of "Daysleeper." And I was pleasantly surprised to find that Madonna's latest song was actually quite entertaining. (*Good for her.*) And the flames and pyrotechnics that Rammstein shot out over the crowd were so beyond what I had ever seen before, I was momentarily concerned for the safety of the audience on the floor. (*Way to kick ass, Rammstein.*) The less said about the rest of the performers the better.

Though what really made the night special was everything that happened *before* the official show started. The night began with a limo picking all of us up at our hotel and taking us to the event. The streets around the Fila Forum were lined with thousands upon thousands of kids. They were all jumping and screaming with the enthusiasm of the kids in the '60's greeting The Beatles as they landed in the States.

They snapped photos of our limo not realizing it was just us inside. Our driver however couldn't resist playing to the crowd. He rolled down his window and yelled, "Everybody! It's the Best Fans! It's the Best Fans!!!" More flashes went off and people even crowded up against the car. Sorry, kids. Michael Stipe and Madonna we were not …

Ultimately, we made it to a private lot. Here we took the time for a brief photo op and then headed for the main doors. This would turn out to be the most stressful few minutes of the entire trip. Everyone was trying to get through those big glass doors. We had to wait while our escort tried to get us past the gatekeepers. The head gatekeeper was a giant specimen of a man who looked a lot like *The Mountain* (*The Hound's* brother on "Game of Thrones"). We had passes that weren't laminates but these big pink puffy things that we hung around our necks or clipped to our belts. So, it should have been easy to distinguish us from the rest of the crowd.

After a couple of minutes the one door opened and The Mountain beckoned us in as the throng of the crowd surged up against us from behind. Chivalry at this point turned out to be a mistake on my part. I paused to allow two of the female contestants in our entourage to enter before us. They got in as people tried to squeeze around us and break through the barrier. Before Julie and I could follow, the door slammed on us to keep out the masses. Oh *crap*... We should have charged the brief opening.

I knocked on the door repeatedly but nothing happened. Minutes passed and I began to wonder if we were going to be left behind and miss the whole thing. Eventually, another brief opportunity presented itself. The Mountain appeared at the door again and I yelled through the glass, "I'm with the *Best Fan* contest!"

He dismissively said, "No, you're not!"

He started to turn away, so I slammed that big pink pass up against the glass and yelled "Yes, *I AM!!!*" Reluctantly, after only a moment's contemplation, he cracked the door once more. No hesitation this time, I immediately lunged through with Julie in tow. But he quickly tried to re-close the door on my arm with Julie still on the other side. I refused to let go and yelled, "She's with me! *She's with me!!!*"

There was no way I was going to be separated from her at that point. I maintained my vise grip on her hand. With no cell phones, I wasn't going to lose her in a foreign country, in a crowd of thousands... Persistence eventually paid off and Julie was finally dragged through another brief opening and the thronging mass was safely blockaded behind us. We left The Mountain behind, and I envision him now doing his best Hodor impression. *Hold the Door! Hold the Door!* For us though, it was on to the party.

We were escorted to the upper level where different corporate sponsors each had their own events taking place around the huge oval concourse. The Carlsberg party was a trip. The beer was flowing and a Beatles tribute band was playing away as people drank, talked and danced. I was delighted to hear music that was right up my alley, and Julie and I shook out the cobwebs of our stressful entrance and kicked back.

Carlsberg employees were weaving through the crowd with mini kegs on their backs in backpacks. They also had holsters on their belts full of cups. As soon as your beer got low or (heaven forbid) *empty*, before you knew it there was someone at your side whipping out another cup from his holster and filling it from his portable keg. "Another Carlsberg for you, sir?"

Suffice it to say we were pleasantly tipsy by showtime. But before we took our seats for the main event, we were told our ceremony was going to take place. We gathered around the Carlsberg stage in the concourse area with the film crews rolling.

I was introduced as the Fan of Honor and as I came up on stage, the cameras panned over to see a goofy-looking, smiling guy with a Marillion shirt proudly displayed on his torso. A friendly wave to the camera was all I got to do—no actual acceptance speech, even though I had one prepared just in case. But getting the Marillion shirt seen was a small victory and getting the Marillion name out there was what this was supposed to be all about. The winner of the Best Music Fan award was announced next and then we were off to see the music acts.

After the show, we once again gained access to the private concourse area where the parties were going to brew all night long. More camera crews were swarming

My brief intro (left) holding a bottle of that night's favorite beer. Check out Julie's expression after the "Best Music Fan" winner raised his trophy (right) and she pulled back so as not to get knocked in the face by it. Also, a Carlsberg attendant can be seen using his backpack keg to fill a beer cup from his holster.

around in an attempt to capture some exclusive peeks at stardom. On two occasions my Marillion *Radiation* shirt was in the camera view while others were being filmed. Then the film crew hired by Carlsberg found me and asked to do a quick interview.

I was told by the interviewer to just look at her, *not the camera*, and answer a few questions. Then a giant camera with an eight-inch lens was in front of me and I stared down its barrel like the proverbial head-lit deer. She asked me about being named the Fan of Honor. I remember replying something on par with, "Uh… Ummm Marillion… uhh internet—uhh fans… umm Tour Fund…" Then before I knew it, the camera was gone and off to someone with something interesting to say. Dear Julie then tried to hide her pitying expression and lied through her teeth as she put a hand on my arm and said, "You did great." *Oh, you sweet little liar, you.*

Well, I definitely blew that. But at least I succeeded once again in getting on camera as a human billboard with my Marillion shirt. We then mingled amid the crowd which included several celebrities (though I never caught a glimpse of Michael Stipe or the R.E.M. guys—the only ones I would have wanted to meet from all the bands that played that night).

We saw a couple of rap stars, a Spice Girl and a few more members from some of these European bands I wasn't familiar with. At one point, a crowd surged around me and I got really annoyed. I didn't realize until I was told later that the guy I was standing right next to was Robbie Williams. *Who?* Yeah. Apparently, he was a big deal at the time. You already knew that? Well, I didn't. But does he know all the words to "Supper's Ready" by Genesis? I didn't think so. Ha. Take *that*, Mr. Williams.

Chapter 19

Acceptance Speech

So, Jeff, if they flew you all the way out to Milan and awarded you with this distinction, how come we didn't see you or Marillion featured in the Carlsberg promos during the MTV awards? Well, the answer to that is a disappointing one. I haven't told the band about this, as I didn't want to throw anyone under the bus, so to speak. But if I'm telling the whole story now, then I guess it's time to finally let this one out of the bag. I apologize in advance for this. I'm really not trying to make anyone look bad. I'm just telling the story.

Just before we were about to embark on this trip, I received a notice from my Carlsberg contact. They told me that they had received a "Letter of Complaint" from someone stating (in so many words) that *they* should be the winner of this prize instead of me. Carlsberg said they didn't want to be involved in any controversy, and so had decided to not feature the Marillion story in their ad campaign before and during the awards show. Our travel was already booked and paid for, so they would still allow us to accompany the other contestants. But we wouldn't be participating in the major promotion of it in any way.

Note that this person never entered the contest. Also, note that I mentioned this person by name in my contest entry giving him credit for the role he played. Furthermore, I clarified my role as the *Idea Man* (as h had dubbed me) when I initially spoke with them on the phone. After that, Marillion further verified with them that I was in fact the person that came up with the tour fund idea (as I had stated in my entry).

I apologized for any misunderstanding and said I was happy to have this person join us and that I'd be more than willing to share the title. I just wanted the Marillion name to get some well-deserved publicity. They responded saying they had already used their budgeted allowance and so couldn't bring any additional people. Sadly, Marillion would not be featured at all. A missed opportunity.

How much of a missed opportunity? A pretty big one it turns out. You see... Once I told them we would be seeing the shows in England before heading to Milan, Carlsberg had planned on sending a local camera crew to the London concert. They were going to feature Marillion in their promos that would be seen by literally *millions* of young music fans tuning in to MTV Europe.

Hi, MTV viewers. Carlsberg here again with another update on the Best Music Fan contest. Look at this crazy kid. He just flew to an entirely different continent—just to see his favorite band play. And not only that, he came up with an idea for all of their fans to pay for that band's tour. And the idea worked! They raised over $60,000 and paid for the whole tour. What a story! He's our Fan of Honour. And that band is Marillion. Now check out this clip of them performing "These Chains" last week at the London Forum—a killer song from their new album Radiation...

<u>A Comically Brief Quote From Marillion Manager Lucy Jordache Upon Reading the Fan of Honor Chapter:</u>

"Jesus!"

Lucy Jordache
July 2021

So yeah. Some serious free promotion left on the table back then. Before they announced the contest winner at the awards show, they thanked me briefly. I don't know if that even made it on the air. I also mentioned being interviewed fleetingly. I don't know if that was ever broadcast either. Basically, I just did my best to get my Marillion T-shirt visible on camera if at all possible.

For what it's worth, I had the following acceptance speech with me just in case. As I re-read it now, I can see things I would have said differently. Yet, I still think it's a pretty good one – brief, yet still heartfelt and with a heavy push to get the name Marillion into the ears of potential new fans. As you know by now though, it never made it out of my jeans pocket that night.

Hello everyone. I want to thank Carlsberg for choosing me as their Fan of Honour. I'm thrilled to be joining them at the MTV Europe Music Awards in Milan.

This distinction may be going to a single person, but this truly was a *group effort*. I'd like to accept this award on behalf of everyone on the Marillion list server ("Freaks") and everyone who donated to the Marillion Tour Fund. As Steve Hogarth said, this was a tour that truly belonged to all of us.

I'd also like to personally thank Rob Crossland & Rob Galardi and Lisa & Christine from Music News Network for their efforts in organizing the Marillion Fan Convention. And of course, Jeff Woods whose efforts are too numerous to convey. The Marillion Tour Fund was brought to fruition because of his tireless dedication and enthusiasm.

What we all did, 1,000+ fans from around the world raising enough money to sponsor a tour for our favorite band may seem crazy to some. Yet any Marillion fan will tell you that it was worth it 1,000 times over. To know the music of Marillion is to be part of something unique and limitless.

If you're a music fan... do yourself a favor. Go out and buy a Marillion CD today. Spend some time with it. Hear what truly innovative music has to offer you if you only give it a chance.

I listen to a lot of music by many different bands, but I always return to Marillion and respond with my heart and soul to what I hear and feel.

Signed,
Jeffrey Pelletier
Music Lover
Carlsberg Fan of Honour

Story 9

OK. That was a fun one, even if the conclusion was a little disappointing. Sorry about that. It's just so frustrating to think about the missed opportunities and wish they had turned out differently. It's probably a good time to get any *non*-happy stories out of the way though, now that we've brought the mood down a bit.

The following is an attempt I made some time ago to speak poetically about death. I didn't think I got where I wanted to with this snapshot, even though I was pleased with some of the imagery. Yet, my Uncle Dan did reply back at the time to tell me how much he loved it. So hopefully it was still a *semi*-success, even if I didn't think it was a full one.

I suppose I should clarify that *yes*, since this writing I *have* eaten pears. Turns out a ripe, juicy pear is really great. Who knew?

Age 4, 1971
"Grandfather the Dragon"

Roger says the land of his fathers is calling to me, and he sits in the canyon with his back to the sea. There's a blood red dragon on a field of green.

I have only two memories of my father's father—one while he was living and one after he was dead. I've heard the stories and I know the facts. He was a devoted alcoholic who oscillated between being ragingly violent and passively pathetic. Yet at the innocent age of four, I had no preconceptions about him.

From the front door, we came directly into the living room with its hard floors and wooden chairs without cushions. The walls and cupboards held so many pictures of angels and saints that stared with sadness at the man slumped at the wooden table. I kept a grip on my father's pant leg because this was foreign territory and my father was the most powerful being in my world.

Dad went into the kitchen leaving me standing on shuffling feet. The creature at the table stirred and in a flash of cognizance became aware of his surroundings. He batted his eyes and peered around at me with the wary apprehension of a dragon awakening from the comforting slumber of sleeping on his hoard of jewels and discovering an intruder in his lair.

The dragon surmised his surroundings; bore his eyes down on me. "Hehhh," he grunted at me.

I took a cautionary step backward. He looked me up and down.

"Hehhh... come 'ere-hhhhh," he said, motioning me toward him.

I stepped back to my original place.

He motioned with more grunts and flapping of his good wing, so I swallowed my fear and paced right up to him.

He turned, reached to the bowl on the table that entertained a fly and pulled a piece from the pile. A flap of the wing chastised the fly. Turning back, the wrinkled dragon proffered his fruit. "Hehhh...," he repeated, but now he was close enough that I felt the fire in the dragon's breath. It blew over me with a sour and sickening smell.

"You wann' a pearhhhh...?" it asked, its head unsteady on its struggling neck.

"No, thank you," I squeaked as Dad returned to the room and my rescue.

Not too long later, the dragon was dead. The aunts and uncles knelt in turn before the open box. Some of them cried. Some of them didn't. Then my brother and I were placed before the casket. "What're we supposed to do?" I asked Alan. He was two years older and thus wiser.

"I don't know. Just sit here a minute," he whispered, expressing his usual annoyance with me.

In the box was the dragon's shell. It looked like it was painted for some reason I didn't understand. A fly buzzed about and was undisturbed when it settled on the dragon. No smoke rose from the dragon's nostrils. No sweat beaded on its brow. No menace emanated from its core. After a long sixty seconds, Alan got up and exited. I followed.

Out in the parking lot, cousin Chris was fidgeting like he couldn't hold his bladder. (This is the same cousin who managed to break something every time he visited, the cost of each disaster increasing every year.) Chris ran right up to us and with sheer delight in his eyes asked, "Did you touch him? Did you *touch* him? He was *cold*!"

Present Day... To this day I have not eaten a pear.

And Chris has not needed therapy. More power to you, Chris.

Story 10

Not to be overly morose, but while we're on the subject of death we also have the following exercise in the examination of mortality. For the record, Danny (the son of Dad's girlfriend, Peggy), did in fact throw a dart so hard at my head that it would have killed me if it had been an inch lower. (Well, it would have at least embedded itself in my skull. I suppose there was a *chance* I could have been one of those miracle survival cases. I'm not counting on that though.) It was an unprompted act by an unstable kid. And I'm just glad I was as short as I was at the time. Here's to not being tall! Even though it sucks at concerts…

Anyway, I present to you another musing on the subject of potential demise, written after another near miss from the reaper.

"Brushes With Death"

Death always seems to be waiting around the corner like a reluctant friend. From the time that one of Dad's girlfriend's sons threw a metal dart so hard at my head that it sunk half an inch deep into the wood cabinet behind me (after skimming through my hair), I've always noticed the calming effect that the threat of actual death seems to bring.

There was the time a flatbed truck pulled out right in front of me in a snowstorm. Pressing the useless brake pedal to the floor I skidded under the bed of the truck, the rear corner of the flatbed smashing through my windshield right in front of my face before my car bounced off the truck's rear tires, rebounding back out from underneath.

Then there was the time back in my high school years when we got bored on that mountain trail we were hiking in the Kancamagus pass and moronically decided to leave the trail and walk up a mountainside. Shuffling across a lazy incline, I slowly started to slide towards the edge and a 100-yard drop, my feet and hands finding no purchase...

And then more recently... I was home sick again with *another* sinus infection. They seemed to be getting worse and worse and coming more frequently. I had just gone to see an allergist for the first time and he was predicting that I had developed a case of allergy-induced asthma. So anyway, I was getting that newly familiar shortness of breath so I knew I had to do something about it. I was worried about driving, so I called around and luckily got hold of my friend Dan Peduzzi (who worked in my hometown). He immediately agreed to come pick me up and drive me to my doctor's clinic.

Thankfully, the nice lady at the front desk brought me right in, 'cause by this point, I was breathing very shallowly and fast and I was starting to feel kind of dizzy. She escorts me back to set me up in one of their exam areas. A nurse looks up from her work, sees me and her expression tells me that something is very wrong.

"Come on," I'm thinking. "I can't look *that* bad, can I? I mean, I know I didn't shower this morning, and I feel a little flush, but it's not *that* big a deal, is it?"

Next thing I know, a heat wave washes over my body and the world starts to tilt. I try to inhale but my chest is so congested that all I hear is gurgling. My throat goes into spasms trying to pull in some air, but it isn't happening. All the energy drains from my body and I collapse, falling forward onto the gurney.

Things start to happen very quickly at this point, but it really *did* seem like it was all in slow motion. The nurse cried for help and with hulk-like strength flipped me onto my back in one quick move. A doctor and two other nurses circled me, but I hadn't been able to breathe for a good ten seconds or so now and that eerie calm had settled in.

The doctor starts belting out the orders. "Get that sweatshirt off of him!"

"I need an O_2 nebulizer!"

My shirt is whipped over my head and something is slapped around my arm. My throat is convulsing faster and I feel so hot ...

A mask is secured on my face.

"Pressure's dropping, pulse is soaring!" the nurse calls out and *wow, she sounds really worried. I wonder why ...*

"Get me 10 cc's of epinephrine, *STAT*!" A body disappears in a white blur.

"What's the oxygen at?"

"Eight!", came the urgent reply.

"Up it to Twelve!"

The mask starts hissing a little louder.

A second later, a needle is plunged into my arm. I hardly notice it.

Despite the medical frenzy, I feel sort of detached and the thoughts in my head are all I can concentrate on.

Wow. I can't breathe. I wonder why I'm not scared. Oh great, they're going to have a wake for me and I hate wakes. I don't want anyone staring at my dead body. Just throw a party with some really good music. Yeah right, fat chance that'll happen.

Urgent activity buzzes around me below the level of my perception.

I wonder if anyone at work will miss me. Sure, it's all they'll talk about for a week or two, 'Did you hear about Jeff? Can you believe it?' But will anyone really think about me six months from now? Maybe a couple of people.

Jeez, is it over already? I never had the chance to do so many things. But, man, these medical people are super efficient. They really seem to know what they're doing too. Well, I hope they make out.

OK. I know a lot of friends are going to miss me. I start naming all the ones who'll *really* miss me—miss me after everyone else has moved on.

Ex-girlfriends? Maybe a few... If they even find out about it, that is...

"Jeff, listen to me."

Oh... the doctor's talking to me. *Hi.*

"Jeff, we just gave you an adrenaline injection. This should help slow down your heart rate."

Adrenaline, slow down? I thought it worked the other way around.

"I want you to try and relax. Just take it easy and try to take deep breaths."

Hey, I'm calm. I wasn't panicking or anything, it was kind of relaxing just letting you guys handle everything. And, hey, I can breathe! And you're right, I'm breathing about three or four times a second here. I'll try to slow down.

"That's better. Take it easy. You're going to be fine."

The next day, telling this story to my friend and fellow engineer Brian Pearson, his reply was "Wow, sounds like a really cool episode of 'ER'."

Chapter 20

The Pre-Order Campaign

The fans considered the Tour Fund a complete success. And they made it known to the band many times on the various internet forums that they were ready and willing to do it all again. Marillion, however, didn't want to want to be the band that's always asking their fans to just give them their money. Consequently, the next two albums to follow 1997's *This Strange Engine* were released under standard distribution deals with independent labels. *Radiation* came out in 1998 and was quickly followed in 1999 by *marillion.com.*

The "io" letters in Radiation were made to appear as the number "10" to signify the band's tenth studio album. The artwork for Dot Com (as the eleventh Marillion album is often called) was another exercise in the band embracing their fanbase. Fans from around the world were asked to mail in passport-style pictures of themselves. These were all cut and physically assembled together into a massive collage of faces from around the world. Marillion's fans can forever see themselves on the slipcase of one of their favorite band's CDs. How cool is that? Regrettably, Julie and I didn't send in pictures because we only had color passport pics and they originally asked for black & white ones. We should have chanced it like so many other fans did (as seen on next page).

The next album wouldn't come out for 2 long years. During that time Marillion held a meeting to discuss their options. They weren't satisfied with the current business model. Yet they didn't want to ask for another Tour Fund. At this point, Mark Kelly came up with an idea.

Instead of asking the fans for money, what if instead they asked them to purchase the album *before* it was recorded? This would give them the cash infusion they needed to pay for the studio time to write and record the album without signing away all the rights to it. And it would feel a lot less like charity.

By this time, Marillion had amassed a massive database of contact info for many of their fans. You see, when they released the *marillion.com* album they included the opportunity to get a free bonus disc in the mail if you simply gave them your contact info. After the Tour Fund, Marillion understood the value of being able to communicate directly and instantly with their fans *en masse*.

Not all of their fans were following them on the list servers. But every single person who bought the *Dot Com* album would see that they could get a free CD if they simply opened up the sleeve to look at the liner notes. And many of them took advantage of that opportunity, providing Marillion with the data they knew would come in handy going forward.

The bonus disc was aptly titled *marillion.co.uk* (using the British internet domain suffix as a companion to the standard *.com* suffix). It included live tracks from their various concert discs as well as an exclusive 9-minute interview and live videos of the songs "Deserve" and "Bell in the Sea."

Now they had a mechanism to reach out instantaneously to an even larger segment of their fanbase.

One example of this database's usefulness has to do with Simon Mayo of BBC radio. In 1999, he made an offhand remark about whether Marillion even existed anymore and further asked, "And who cares anyway?" Marillion immediately contacted their fans with the BBC e-mail address and suggested they all let the BBC know that Marillion was in fact still alive and kicking. The response overloaded the BBC servers. Eventually, the BBC cried "Uncle" and even apologized for the dismissive comment and interviewed Mark Kelly on BBC radio.

In early 2000, Lucy Jordache (formerly of EMI and now Marillion's full-time manager) sent a message to their fans. She explained the pre-order plan and asked them, would they be willing to buy the album a year before it was made. Before any of the music had even been written? The response was better than they expected.

They had estimated that this plan could work if they pre-sold about 8,000 copies of the unwritten, unnamed next Marillion album. Within a few weeks, they exceeded that number and raised over £100,000. In the end, more than 12,000 fans pre-ordered the 12th Marillion CD a full year in advance of its release.

When asked about this by Prog magazine, Ian Mosley recalls the following:

"I remember we were getting ready to start work on what would become Anoraknophobia. And it was Mark that said, 'Let's email all the fans and see what they think about this idea of giving us some money upfront before we've written or recorded a note.' All I could think was, it's worth a shot. We had about twelve or thirteen thousand fans, and they donated money. People from all around the world, which was unbelievable. It really felt like we had the freedom to do what we wanted again."

And thus, the first Pre-Order campaign was born. As with all great ideas, this quickly became commonplace in the music industry. So many bands now launch pre-order campaigns for their musical output. Yet, before Marillion did it, it had never been done before.

Chapter 21

Intellectual Property

No patents were ever issued for the Tour Fund. However, if they *had* been, the Pre-Order campaign would have been considered a "New Use" under Intellectual Property law. If you improve on an existing patent or find a previously undiscovered new use, then you yourself have also invented something. And so, I very proudly consider Mark Kelly to be a *Co-Inventor* of the modern crowdfunding business model. He invented the Album Pre-Order model.

Feel free to listen to Mark's TED talk on "The Birth of Crowdfunding" if you can't get enough of this stuff. If you don't have the patience to search for it, you can find it here: http://tinyurl.com/MarksTED

Speaking of intellectual property, do you mind if we jump ahead in our timeline briefly? No? Good.

So in 2013, my phone rang one more time with another of those unexpected calls. This time it was the legal team for a major crowdfunding company asking if I was the Jeff Pelletier who created the Marillion Tour Fund. It turns out they were in a legal dispute based on someone claiming ownership of some of the crowdfunding intellectual property based on patents issued in the early 2000's.

I was hired as an expert witness to provide documentation that the Marillion Tour Fund (the original crowdfunding enterprise) pre-dated those patents by over 5 years. I also made sure they were in contact with Mark Kelly (which they already were). He was also hired to do the same thing regarding the Pre-Order campaign, which pre-dated the patents in question as well. This was a nice exercise for the two of us. Too bad it didn't result in us being given an ownership share of the business model that we created.

And now back to the Pre-Order. (See? I said that would be brief.)

In September of 2000, Marillion issued the following Press Release regarding the Pre-Order:

Anoraknophobia Pre-Order Press Release

Music Industry First as Marillion Fans Finance New Album, September 2000

In what is believed to be an unprecedented venture which could revolutionize the music industry, Marillion fans have financed the making of the band's latest album after an Internet whip-round raised over £100,000 - conclusive proof of the power of the Internet for artists and record companies alike.

Marillion, who have been writing and recording for almost 20 years, approached their database of over 30,000 fans via e-mail, as an alternative to taking up the deals they were being offered from established record companies. The response was overwhelmingly positive: within 3 weeks fans offered to pre-pay for the album, to the tune of 5% of total world-wide expected sales, providing enough money to cover the costs of making their 12th studio album. In doing so, a record company advance was rendered redundant and the band retained the rights to their new music.

This groundbreaking idea enabled the band to return to EMI Records who have agreed to license the album for a world-wide marketing and distribution deal. The album will be released in Spring 2001 through their Liberty label.

Any pre-orders will be manufactured and fulfilled by Marillion's own Internet mail-order company, Racket Records, and shipped on the week of release. Fans pre-ordering albums within an initial limited period will be thanked personally in the sleeve notes of the special first edition of the album. Marillion will cease manufacture and sale of the pre-ordered version upon the album release after which the album will be available only at retail.

Commenting on the initiative, Lucy Jordache, Marillion's marketing manager said: "We were being offered deals from various record companies, but what the guys really wanted, was to have total control of their music, yet still be able to utilize the expertise and distribution facilities of a major record label. This could only be achieved if we obtained the capital to record an album from another source, and then took the finished product to a label. This is a real testament to the loyal support of Marillion's fans, old and new. It also demonstrates the power of the Internet and what it has to offer both artists and the record industry as a whole. 95% of the band's market still remains beyond the scope of the pre-order idea, so retailers should see no noticeable loss in potential sales. Indeed, an upturn is expected as Marillion return to their original EMI stable."

EMI Liberty's co-director Peter Duckworth added: "We were very impressed with this venture which we believe breaks new ground in the industry. We are all in a win, win situation, EMI are happy, retail is happy, the band are happy and the fans are extremely happy."

Marillion are no strangers to harnessing the power of the Internet. They were one of the first groups in the UK to set up a web-site, ("http://www.marillion.com"), in the mid 90's to communicate with and sell Marillion products to their fans - fans from as far and wide as Brazil to Japan and Australia to Iceland. They now have an active database of over 30,000 fans who visit their web-site on an average weekly basis.

UK fans objecting to a chance remark by Simon Mayo on Radio 1 last summer, brought the radio station's computer system to a standstill when they emailed to complain and demanded he play a Marillion track. He subsequently interviewed keyboard-player Mark Kelly on air in an attempt to understand and appease the phenomenal passion of Marillion's audience.

Back in '97 American fans underwrote an entire US tour to the tune of $60,000, with donations following an internet campaign - an idea conceived and managed by the fans before any involvement by the band.

Lead singer, Steve "h" Hogarth adds: "It's not just about the money - the Internet allows us to communicate directly to our fans world-wide in a way that's spontaneous and instant. It's a two-way communication process that's changed everything for us - the fans feel like a world-wide family now. Faith moves mountains so watch out."

Chapter 22

A Band of Firsts

It's surprising how many firsts are floating around in this Marillion-verse. Let's review what we already know.

- The Tour Fund was the first successful internet-based crowdfunding campaign.
- The North American *This Strange Engine* tour was the first fully fan-funded tour of a major record label artist without money from the music industry.
- *MARILLIONROCHESTER* (The Tour Fund CD) was the first product of a crowdfunding campaign.
- *Anoraknophobia* was the product of the first CD Pre-Order campaign.
- And although not verifiably the *first*, Marillion was *one of the first* bands in the UK to set up a website in the 1990's—always at the forefront of technological advancements.

That's pretty impressive. Are there more firsts though? Well… Did you know that Marillion was also the first major band to use wireless MIDI technology in their performance? MIDI stands for Musical Instrument Digital Interface. It allows you to store samples of music that can be played back on command.

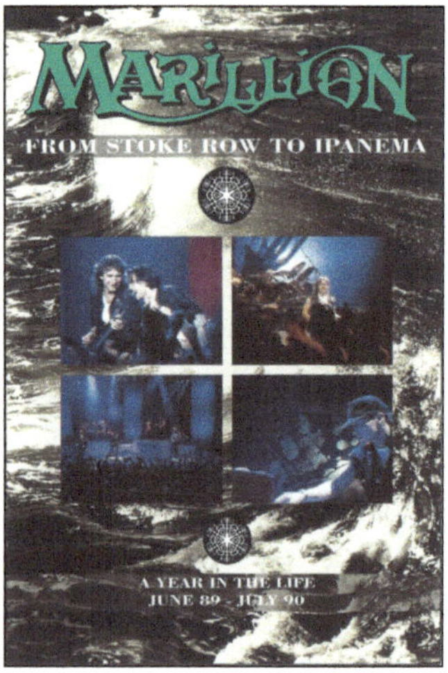

In the *From Stoke Row to Ipanema* DVD, h discusses how he contacted John Price, an expert in MIDI technology. The goal was for h to be able to play samples in a completely wireless format so he wouldn't be "tied down" during his performance. Together they developed a pair of gloves with switches in the fingertips that would trigger these samples. During the *Seasons End* tour, h played back these samples using the MIDI gloves by tapping his fingers against a large vertical pane of glass on stage. This can also be seen in the *Stoke Row* documentary video (released in 1990).

These MIDI switches were later incorporated into a cricket bat that h would play on stage for many years to follow—most often serving as the method to provide the sax solo in live renditions of the song "This Strange Engine."

Did you know that Marillion is also in the *Guinness Book of World Records*? In 2003 Marillion successfully broke the record for "fastest concert video release" by recording the Friday night *Afraid of Sunlight* performance at their 2003 convention in Minehead, England. The video editing happened on the fly (while the concert was still happening) and throughout the night. The *Before First Light* DVD was immediately printed and was available for purchase at the convention that very same weekend, and then in stores the day after that.

Once again, Marillion showed their never-ending appreciation of the devotion of their fanbase. If you watch the credits on this DVD you will see all 2,000 names of everyone in attendance that night scroll up your screen for over four minutes while h sings the demo version of the title track to *Afraid of Sunlight*. This is humorously prefaced by the line—"And introducing as The Audience..."

On April 1, 2003, Marillion issued the following press release detailing the record-breaking event and the mad dash to a physical record store on Monday to finally secure the official time.

1 April 2003

Marillion Set New World Record for Music DVD

On Monday March 17th 2003, Marillion accomplished a record-setting world-first at their 2003 fan convention.

A mere 63 hours after Marillion walked on stage to perform their 1995 album, Afraid of Sunlight, a full DVD of the concert went on sale—making it the fastest DVD filming and production in history.

The clock began ticking the minute the eight cameras began to roll on the Friday evening. Less than 48 hours later, "Before First Light" was available for sale at the convention to the 2,000 fans who'd come from around the world.

However, in order to qualify for a Guinness World Record, the DVD had to be available in a traditional retail outlet, so, after a frantic dash the official record pegged at 63 hours when it went on sale in The Record House, High Street, Aylesbury, Bucks. The DVD is also on sale at www.marillion.com

The official Guinness World Record listing reads: 'The fastest time for a music DVD to be filmed and then released is 63 hr 29 min. Featuring the band Marillion, it was recorded at the Marillion Convention at Butlins, Minehead, Somerset UK on 14 March 2003, before going on sale to the public on 17 March 2003.'

Marillion have what is occasionally described as a fanatical fan-base that have supported them through more than two decades of prolific music-making. The weekend convention was their second sold-out international fan gathering with people traveling from as far away as North and South America, Australia, Poland and Mexico.

Since Guinness used the *later* date of the first in-store sale for the timing of the official record (instead of the earlier sales at the convention), Marillion repeated the whole exercise in 2013 at the Wolverhampton Civic Arena. They did it even faster the second time around, and successfully got the record time reduced with the *Clocks Already Ticking* DVD. That record stands as of this writing. These two concerts are now available together on the *Breaking Records* 2-Disc Blu-Ray release.

Guinness has the following entry listed on its world records website. (Anyone up for a "Weekender?")

The fastest music DVD (filming to release) is British band Marillion's DVD entitled 'Clocks Already Ticking', which was filmed, edited, and duplicated for sale in 10 hrs 31 mins. The DVD was filmed at the "Marillion Weekender" event at Wolverhampton Civic Hall in Wolverhampton, UK, from 12-13 April 2013. Marillion previously held this record in 2003. For this attempt, the DVD went on sale at the venue and online at the "Sandbag" music retail outlet. The DVD was produced by Abbey Road Studios, Toward Infinity and Racket Records.

Chapter 23

Racket Fuel and Rubber Soul

On November 16 & 17, 2000, Marillion played two gigs at the Bass Brewery Museum in Burton-on-Trent, England. To commemorate the event, Bass brewed up a special batch of Marillion-themed beer. It was called "Racket Fuel." Marillion's own recording studio was called "The Racket Club." And the name was a nice twist on actual rocket fuel.

I didn't attend those concerts, but I had the pleasure of sampling the brew thanks to the generosity of long-time Marillion friend, Jeff Schwartz. Not only was he kind enough to share his bottle of Racket Fuel; but before I could protest for him not to, he proceeded to remove the label from the bottle, tear it in half and present me with one of those bisected pieces. (*See below*). Next drink's on me, Jeff!

Marillion. Established 1982. And now in bottles!

The concert was originally only available as an online webcast. However, Marillion issued a free CD of the show to every member of their fan club as their 2000 Christmas gift. The show was called *A Piss-Up in a Brewery*. After they determined that there was indeed sufficient demand to justify an official video release, a full video of the same name was released on DVD in 2002.

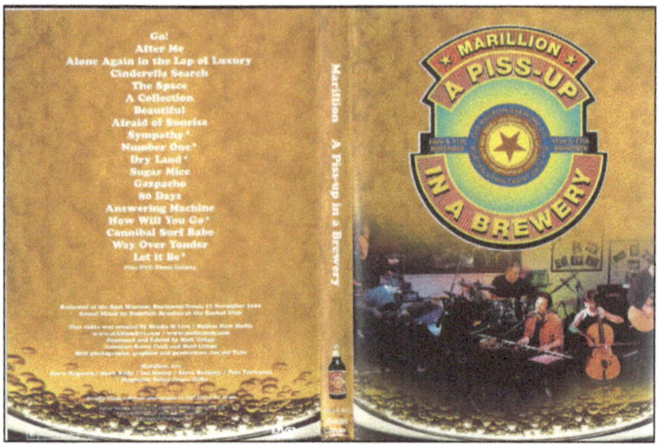

So that adds to our tally of beer connections in our Marillion story. Just for fun, let's recap them. There is only one degree of separation between Marillion and:

- Carlsberg beer: (the Fan of Honor Award)
- Guinness beer: (the World Record video release)
- Bass beer: (the concert at their brewery museum)
- Heineken beer: (they played the Heineken Music Hall, Amsterdam in May 2001)

That last one is a bit tenuous. But if we're going to play the game, a tenuous connection is still a valid answer in filling out the *Scattergory*. Did I miss any? Let me know if I did.

Marillion also had a line of beer-related merchandise in the form of T-shirts, (artwork seen above). And (on a personal "Marillion & beer" connection), we toured the Heineken brewery on one of our Marillion trips to Holland. A good time was had by all —as seen in the animated snap above...

So, you may or may not find this interesting, but Julie and I have experienced a wonderful interweaving of The Beatles and Marillion during our travels. Just for kicks, here's a quick review...

- In 1997, during the Tour Fund convention, Marillion played a cover of "You've Got to Hide Your Love Away," while we sat on the edge of the stage. Plus we discussed the "Penny Lane"—"80 Days" connection with Pete.

- In 1998, during the Carlsberg party at the MTV awards in Milan, the Beatles cover band played the pre-awards festivities.

Channeling Ringo and the drums in "The End" (left) at the Carlsberg party. And belting out "I've Got a Feeling" into an unplugged mic (right).

- In 2000, during the Marillion Xmas tour, we saw two shows in Leeds and London and visited the Beatles museum while in London.
- In 2002 or 2003, Julie and I spent the evening at a pub around the corner from our hotel in London. There, a delightful old man was playing piano and the crowd was singing Beatles songs with him all night. I cannot overstate how much fun we had that evening—especially a delightfully tipsy Julie.
- In 2007, at the first Port Zélande convention, h joined a club band at an after-show party and started singing Beatles covers (including "Helter Skelter").
- Again in 2007, before the Port Zélande convention, we toured the gorgeous John Lennon honeymoon suite while staying at the famous Amsterdam Hilton, immortalized in the song "The Ballad of John & Yoko." They don't even advertise this room—at least they didn't back then. If you ask though, they will let you

"Drove from Paris to the Amsterdam Hilton, talking in our beds for a week. The news people said, 'Say what you doing in bed?' I said we're only trying to get us some peace."

see it at no charge (still, a tip to the bellman is greatly appreciated, I'm sure). As you walk around the room, you can see lyrics stenciled on the walls and sense that palpable energy from standing in a place of historical significance.

- And finally, at one of the Montreal conventions in the 2010's, we went to a John Lennon exhibit that Yoko Ono had commissioned. On site was the famous white piano from "Imagine." I have to say, it gave me chills as Julie and I were allowed to sit before this historic instrument. With a single hand, I tapped out the opening seven notes to that simple, yet incredible song…
- Given my luck, I truly expect to run into Paul McCartney at a coffee shop on our next Marillion trip. I can't wait to talk to him about music over tasty white chocolate lattes. I fully intend (in homage to Chris Farley) to bring up that line in "The End"—*And in the end, the love you take, is equal to the love, you make.* I will then ask, *Is that true?*

A voice from the past, entered my head today
Fresh and alive, full of life, passion and pain
A voice now past, a beautiful soul. Gone, gone, gone
... talking perfect sense to the next generation.

- Lyrics by Steve Hogarth

OK. Maybe not relevant to the narrative that we've established so far. But I like the coincidence and I love The Beatles. So what was the harm in that? Now back to the '70's to talk about… what else? Baseball, of course.

Story 11

The following was written in the middle of the Pedro Martinez era in New England—a time when I couldn't help but be enthralled by the greatest pitcher I've ever seen, on a team before they overcame their near-century-long championship drought. Believe me when I tell you that on the days that Pedro pitched, it wasn't just a game. It was an *event*. And that unifying struggle of never quite seeing our team achieve the ultimate prize bonded friends and strangers alike in Boston as we kept saying our mantra, again and again, season after season… *This could be the year.*

But this story isn't about Pedro—even though I could talk about him for hours. Who remembers the '95 playoffs against Cleveland? When in Game 5, an injured Pedro sat in the bullpen while the camera zoomed in and lingered on his cleats? We and the announcers talking about how those cleats (as opposed to regular shoes) meant he might actually *pitch in relief*. And how he did just that and shut down the powerhouse Indians lineup for six straight scoreless relief innings in the middle of a slugfest to clinch the series. That may be the best performance of his career, despite how we later learned that his injured arm couldn't deliver his usual fastball. So he did it all with curves and offspeed stuff. Amazing.

No, it's not about Pedro, or even the Red Sox specifically. This is about baseball in general—a sport that is derided by some but loved by many. Baseball has always been a part of my life, for as long as I can remember. It can bring friends and family closer and it can give strangers something to talk about with each other ... and maybe help turn those strangers into friends. (Hey, Kim Robinson. How 'bout that 2004 Yankees-Sox playoff series?)

My dad actually coached the Little League team that both my brother and I played

Ever the athletic child, he says sarcastically. Did I even know it was a football I was holding (left)? Kind of like giving a monkey a calculator, I'm guessing.

on. I was not a good player. I sucked in fact. And yet I still have nothing but wonderful memories about those games we played together, even though I probably had the worst batting average in the league.

My Mom took us to Fenway Park every year for our birthdays. Those annual trips to such sacred ground are held in my memory as the embodiment of childhood happiness. Baseball is tied to my youth and part of my soul.

Age 8, 1975
"Why I Love Baseball"

Fenway Park in the 70's (left)—one of my first Red Sox games. And on the hallowed field at the first exclusive Father's Day Catch (right) with Julie in the early 2000's.

With the Red Sox surging as of late and pennant fever sweeping across New England once again, it seemed like an appropriate time to talk about our national pastime and why it means so much to so many of us (particularly me).

It's not just the technical things, like the classic one-on-one battle between pitcher and batter. Or the many layers of subtle strategy that take place on a team level. Or how baseball is the only sport where the season actually *means* something. (Look at hockey and basketball where *more than half* the teams get to go to the playoffs. How can you even take that seriously? When a team with a losing record gets to play in the postseason?) Or how, unlike *timed* sports, the game is never over until the last out is made. Any baseball game can be won by either team no matter how far they may be behind. (How anticlimactic is it to see a football player taking a snap and a non-competitive knee to let the clock run out? I mean, I know it's the safe thing to do. But don't you sometimes want to just say, "Come on! Wrap both your arms around the ball and *run* with it at least, and see if you can make another score.")

No, it's more than that. It's that baseball is tied to my childhood. Every year, Mom would let both Alan and me choose a game to go to for our birthdays. They always had a game within a few days of your special day (as long as you weren't born in the winter, which we weren't). I tell you, there's nothing like that feeling when you're a little kid and you walk up from the darkness of the concourse into the sun to take your seats at Fenway Park and you catch that first glimpse of the field. It's pure magic.

These are the guys you watch play on TV every day. And they're right in front of you. As you take in the stunning panoramic view, there are these guys walking around yelling out "Peanuts!" and "Fenway Franks!" and "Get your Cracker Jacks here!"

The Fenway Franks were great. I always wanted to eat one more hot dog than the last year to prove how much I was growing. (I think four was the record, and I'm pretty sure I regretted the fourth one.)

In the Fall of 1975, I was eight years old and the Red Sox were in the World Series. I knew it was something special, but at that age, I didn't have the perspective to realize just how rare an occasion this was. Alan had obtained special permission to stay up late and watch the games. And if he got to stay up late, a little complaining, and I was there too!

I remember sacking out in front of the TV with our sleeping bags and our homemade scorecards. This was as good as it got. Of course (never having stayed up that late before) I fell asleep midway through each game, but one night I woke to the shakes and screams of my brother. I bolted out of the grogginess of sleep instantly (like only a kid can do), to see an indisputable Historic Moment playing out before my eyes... Carlton Fisk waving his arms and the announcer saying, "...if it stays fair, it's out of here!" Alan and I were on our feet at this point. And the ball hits the foul pole and it's a HOME RUN!!! Alan and I laughing and jumping for joy, screaming with glee in what seemed like the middle of the night.

That was a great memory of watching baseball. Here's another one of actually *playing*.

The Andover White Sox of 1978 with tiny Jeff (center) up at the plate wearing Carl Yastrzemski's #8.

Little League is generally a nightmare. Let's face it, ten-year-old kids just don't have the coordination to really play the game with any talent or consistency. Our team, the Andover White Sox, was certainly no exception. To emphasize the point, our outfield did not catch a *single* fly ball all season. Not one. They all hit the ground unimpeded or clanked off a glove or were bobbled on their inevitable way to touch grass.

I was a short kid, and I mean *really* short. I was so short (*How short was he!?*)... so short that a pitch that bounced off the plate was called a strike on me not *once*, but on *two* separate occasions. I shit you not. Regardless, with my microscopic strike zone, my specialty was the "base on balls." I imagined that my on-base percentage was pretty good even if I didn't get that many hits. I can't say that was actually the case, just that at the time I imagined it to be so.

Well, it was toward the end of the season, and believe it or not, we were actually going to win this game if we got one more out. I was playing second base and participating in the usual chatter. All Little League coaches love chatter. Don't ask me why.

"Hey, batter, batter, *swing* batter!"

"Strike him out!"

"He's a whiffer!"

Come on, hit it to me, I was thinking. I'd never been the hero and Little League ground balls were pretty slow. And there's a good chance you won't screw up a toss to first base when the guy's standing only 10 feet away from you. So I'm looking for a slow roller right at me so I can make the final out.

Here's the pitch and then the crack of the bat and the ball takes off into the sky. It keeps going higher and higher... (Accept a little bit of embellishment here due to nostalgic perspective.) This ball was so high it was going into orbit. Everyone's staring at the sky except for the baserunners who are making their rounds. Finally, the ball reaches its pinnacle and starts to drop back to earth.

At this point, it dawns on me that the ball is heading toward *me*. I raise my glove hesitantly and take a few steps back. I can hear my teammates cheering me on, "Come on! *Catch* it, Jeff!"

And this ball is screaming down out of the heavens now and I back up a few more steps. But now I hear my father's voice in my mind marveling at the fact that no fly ball to our outfield has ever been caught and I realize that I've backed up just enough to be standing on the grass instead of the infield dirt.

Oh no! I'm in the outfield. I'll never catch it!

Those of you who've seen the movie "Parenthood" with Steve Martin know what I mean when I say that I must have had my "tense face" on. I take two steps to the left and then back to the right, and the ball's coming down and I'm trying to stay under it.

Oh crap, why'd I wish he'd hit it to me! I'll be humiliated.

And the ball is still coming down faster and faster. Another step to the left—I think I'm going to be sick—and finally, I put the glove between my face and the ball and close my eyes tight and SMACK! The ball smashes into my glove and—*Oww! That hurts!*—I close my glove and MAKE THE CATCH!!!

Gloves and baseball caps are thrown in the air. Screams and cheers erupt on the field and in the stands. And everyone runs over and I'm the center of the mob scene of kids. I'd never been happier.

Every kid should get to be the hero on the field of play at least once in his life. I felt it once and only once, ever so briefly in a game of no consequence. But that fleeting moment shines in my memory with a bright radiance that warms me to this day.

And that's why I love baseball.

Story 12

"The Women in My Life"

I think I've done a pretty good job of not suffering from sexual ambiguity, even though the cards were stacked against me from the start. Mom wanted girls. Alan came first, so they tried again. But the Pelletier Y-chromosomes were just too damn assertive. Regrettably, I was another male. After me, they gave up and I think Mom just decided to pretend I was a girl anyway. We won't even get *into* the pictures I've seen of me in those damn *knee-socks.*

I suppose I have no one to blame but myself for the humiliation I would occasionally be subjected to whenever we went out. I was just a normal kid who hated haircuts, even though my hair had a tendency to curl up when it got below my ears. Waitresses would sometimes comment to my mother, "What an *adorable* little girl you have, Ma'am."

To which I would cry out in frustration, "I'm a BOY!" and then slide down in my chair in embarrassment.

I suppose if I had any kind of sense of humor about the subject at the time, I could've relished the awkward stammering of the confused, apologetic waitress. But no, this was no laughing matter to a six-year old trying to assert his masculinity.

Mom would then ruffle my hair and say something like, "Well, Peanut, if you'd let me cut your hair shorter this wouldn't happen."

"I look like a *goober* with short hair," was my disgruntled reply as I would shy away from the tender hand.

Then, of course, Grandma would pipe up in her disapproving tone, "When are they going to get *boy's* haircuts?"

"This *is* a boy's haircut!" I would bellow.

"Not in *my* day, it wasn't," was the age-old observation.

Then I'd pout until the food came and I forgot about the incident, sipping on my Shirley Temple.

I guess it's no surprise that I keep my hair short now and cultivate facial hair as if I still need to set the record straight. And I suppose it's also not much of a curiosity that most of my best friends have been female.

The male I spent the most time with (my older brother) was my tormentor, whose life's mission was to vent upon me all of his anger and antagonism. He was a cauldron of frustration with this cruel world that must have dealt him some unspeakable outrage–one that I assume coincided exactly with my birth.

And the female I spent the most time with (my mother of course) was still my nurturer and protector back then – the temporary sanctuary from the fear.

Given those two facts, a young mind could easily develop the hypothesis that *"Male = Bad* and *Female = Good."*

This isn't to say I haven't reveled in the male-bonding rituals in my time, but haven't you noticed there's something so inherently enjoyable about having close friends of the opposite sex?

Let's just say, here's to the women in my life. I love all of you, but am so grateful I'm not one of you.

So, what's the point? I don't know. I suppose there isn't one. But there wasn't much of a point in me writing about a spinning, crapping dog either, was there?

Chapter 24

Anoraknophobia

The first-ever pre-order campaign was for the Marillion album later to be titled *Anoraknophobia*. In British slang, an anorak is a compulsive person who focuses on a specific obsession. It came to be because of the anorak hoodie that trainspotters would wear. Arachnophobia is a fear of spiders and is also the title of a popular horror film. Thus, the title of the album has a familiar sound to it, but with a clever spin in referencing the obsessive nature of Marillion's fans.

Once again Marillion showed their appreciation to those diehard fans. The Pre-Order Special Edition of the *Anoraknophobia* CD had expanded packaging and a bo-

---

Hidden amongst those thousands and thousands of printed names was the evidence of Marillion's collective sense of humor. If you look closely at the list, you'll find quite a few interesting (some famous and some completely fictional) names.

I'm sure there's more, but I found the following names during my own casual perusals of the list: Ian Anderson, Bilbo Baggins, Barry (the album's hooded cartoon mascot), Ichabod Crane, Arthur Dent, Frank Drebin, Tyler Durden, Barton Fink, Axel Foley, Robert Frost, Wot Gorilla, Sherlock Holmes, Waring Hudsucker, Clark Kent, Obi-Wan Kenobi, Verbal Kint, James T. Kirk, King Kong, Lois Lane, Jeffrey Lebowski, Lord Lucan, John Merrick, Fox Mulder, Jean-Luc Picard, Harry Potter, Austin Powers, Ford Prefect, Dana Scully, Homer Simpson, Luke Skywalker, Han Solo, Keyser Soze, Egon Spengler, Ray Stantz, Uncle Walter, Darth Vader, Peter Venkman, Bruce Wayne, Steve Wilson, Willy Wonka and Winstone Zedmore [sic].

Many of you will recognize names from the likes of Star Wars, Star Trek, The Usual Suspects, Ghostbusters, Hitchhiker's Guide to the Galaxy, Fight Club, The Hobbit, The X-Files, multiple Coen brothers' films and more. Good one, Marillion.

---

nus disc with extra songs and video. The special edition package thanked over 12,000 people by name on over a dozen pages. Every single person who pre-ordered the CD could see their own names in the liner notes.

In his new autobiography, "Marillion Misadventures & Marathons–The Life & Times of Mad Jack" (seen below with the excellent cover photo by Anne-Marie Forker), Mark Kelly had this to say about the Anoraknophobia pre-order:

"It sounds like small beer these days, but was pretty revolutionary at the time, as was deluxe packaging and an array of formats. Although Marillion have been credited as the inventors of online crowdfunding ... and I sometimes get credited as the band member who came up with the idea, it doesn't take a genius to see how it evolved from Jeff Pelletier's tour fund idea. We just ran with it. Besides, crowdfunding as a concept had been around in one form or another since at least the 19th century. Remember the debentures at the Royal Albert Hall?"

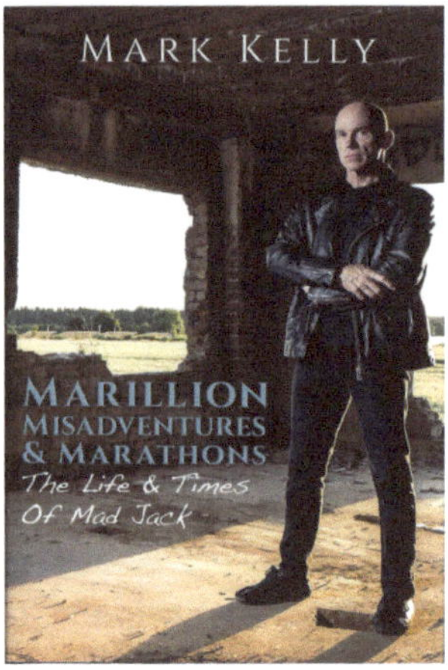

Thanks for the credit, Mark. And thanks for the quotes and the permission to use them. Much appreciated!

In case you don't have this historic CD (and can't make out what's written in the picture of the liner notes on the next page), Marillion (in particular Steve Hogarth) had this to say to all of us on the pages within:

The majority of copies of Anoraknophobia will be bought in record shops. This will be the standard CD on EMI Liberty (and Sanctuary in North America). What you have here is strictly for the anoraks– strictly for "family."

This is a revolutionary record. When we first had the idea to throw the recording contracts in the bin and come straight to you to finance the recording of Anoraknophobia, to be honest, I had every faith that you would trust us with your money. However, I thought that our target of 10,000 presales was wildly optimistic. Privately, we all crossed our fingers and hoped for 8,000. After the very positive response in the first months, I think the band felt a certain unspoken moral pressure to make a CD worthy of your faith and commitment. With Dave Meegan back in the producer's chair we felt we had all the right ingredients in the bowl. Although, as usual, none of us knew what kind of cake we were baking. Now it's finished and the five of us are very happy with what you are about to hear.

This isn't just a record – it's your record. You bought it before we made it. If you were one of the first to pledge your faith, then your name is in this package, and your spirit is in this record. We've tried to make this CD as special as we can in every way. We've added a free bonus disc which contains video of some of the recording sessions, along with rough mixes of work in progress. Mark Kelly's radical remix of "Between You and Me," and an additional new song "Number One" which – if you have a computer – you can actually see being recorded.

We'd like to thank THE boom boom BOYS for editing and creating the video. I should also mention the fabulous Carl Glover who designed the special package.

Most of all, Marillion would like to thank you for giving us your money all those months ago. In the end, 12,674 of you bought this album before it existed. That's a serious hardcore family... and it's growing.

This is the 21st century and everything is hype – but not this. This is about passion.

It's good to be an anorak. We're anoraks too.

See you on tour. (wear anything you like)

Love

h

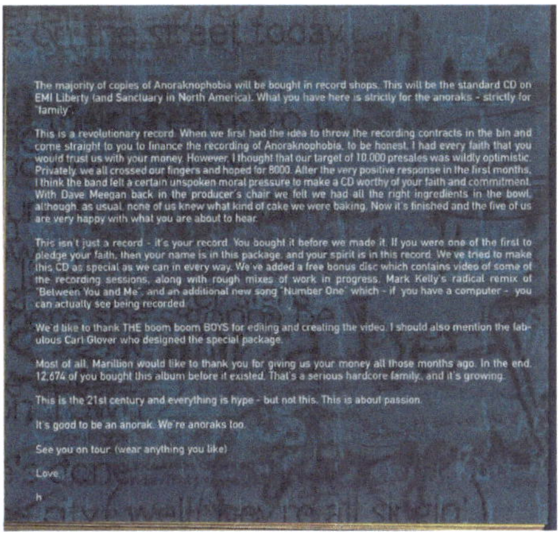

The majority of copies of Anoraknophobia will be bought in record shops. This will be the standard CD on EMI Liberty (and Sanctuary in North America). What you have here is strictly for the anoraks - strictly for "family".

This is a revolutionary record. When we first had the idea to throw the recording contracts in the bin and come straight to you to finance the recording of Anoraknophobia, to be honest, I had every faith that you would trust us with your money. However, I thought that our target of 10,000 presales was wildly optimistic. Privately, we all crossed our fingers and hoped for 8000. After the very positive response in the first months, I think the band felt a certain unspoken moral pressure to make a CD worthy of your faith and commitment. With Dave Meegan back in the producer's chair we felt we had all the right ingredients in the bowl, although, as usual, none of us knew what kind of cake we were baking. Now it's finished and the five of us are very happy with what you are about to hear.

This isn't just a record - it's your record. You bought it before we made it. If you were one of the first to pledge your faith, then your name is in this package, and your spirit is in this record. We've tried to make this CD as special as we can in every way. We've added a free bonus disc which contains video of some of the recording sessions, along with rough mixes of work in progress. Mark Kelly's radical remix of "Between You and Me", and an additional new song "Number One" which - if you have a computer - you can actually see being recorded.

We'd like to thank THE boom boom BOYS for editing and creating the video. I should also mention the fabulous Carl Glover who designed the special package.

Most of all, Marillion would like to thank you for giving us your money all those months ago. In the end, 12,674 of you bought this album before it existed. That's a serious hardcore family.. and it's growing.

This is the 21st century and everything is hype - but not this. This is about passion.

It's good to be an anorak. We're anoraks too.

See you on tour. (wear anything you like)

Love

h

Now fully realizing the power of information, Marillion offered to send a free sampler CD to anyone that would simply provide their contact info via Marillion's website. The CD was called *Crash Course* and can be seen here.

A couple of years later they updated the *Crash Course* CD and continued to make the same offer to any fan (or potential new fan) who wanted to give their music a chance.

The free *Crash Course* offer continues to this day. The pull-down menu for Music on Marillion's website contains a link to *Crash Course*. Currently, you can download a sampler CD full of Marillion music instead of a physical disc. As always, all you have to do is give them a valid email address. This is what they have to say about the *Crash Course* offer:

CRASH COURSE FREE DOWNLOAD

Marillion are used to fighting misconceptions.
Marillion are not what you think they are.
No, really, they're not.
Just remember that whatever your pre-conceptions are about this band, chances are they are wrong.
Listen.
Find out for yourself.
Make up your own mind ...
We have created "Crash Course", an album-length sampler to introduce you to exactly what we do.
If you're new to Marillion, you can download this sampler album for FREE!

Rocking Out 4

Over the years I've written reviews for various publications. Don't worry. I won't force you to read all of them (or even make you have to skip past them). We'll just look at this one since it pertains to the subject currently at hand in our story.

At the time the Pre-Order CD came out, I was asked to submit a review for the album. Here is what I wrote at the time, which I hope stands up well today. Once again, feel free to skip ahead if you prefer. It's your book now.

Age 33, 2001
Anoraknophobia Album Review

Marillion – Anoraknophobia

Released in the US May 14, 2001

Sanctuary Records—Distributed by BMG

Review by Jeffrey Pelletier

Is there ever a better feeling than the anticipation we feel when we've got a brand-new CD by our favorite band in our hands? Tearing it open while making a beeline for the CD player? That band is **Marillion** for me and the new CD is *Anoraknophobia*—a disc of historical significance because of the hugely successful pre-order campaign that took place last year. Over 12,000 fans from around the world paid for this CD over 6 months in advance—before a single chord had been written or recorded, opening up doors for Marillion that were closed to them before.

First off—what's up with the name? North American fans might not realize that an "anorak" is one of those hooded, pull-over jackets that in the U.K. (because of its ties to obsessive "trainspotters") has evolved into a term for anyone with a passion—a term that could aptly be applied to most Marillion fans. But fear not. The title breaks down as follows: anorak-no-phobia. In other words, no fear of the anorak. As (lead singer) "h" states, they really just wanted to say, "We know you're all anoraks and there's nothing wrong with it! We're anoraks too."

If you pre-ordered the CD you knew two things going in. You'd be thanked in the liner notes and you'd get a free bonus CD to accompany the official one. Expectations could be pretty high under such circumstances, but Marillion have outdone themselves once again. This limited-edition package is literally a hardcover book with pages of information including, of course, over 12,000 names. (Look closely and you may find quite a few names that you'll recognize...) It's truly a wonderful package that I'm going to treasure for a long time.

The music itself is another evolutionary step forward in the ever-challenging dynamic that is Marillion. This album has a definite groove and sounds worlds apart from the Marillion we've previously known. The rhythm section is frequently in the front of the mix and bassist Pete Trewavas is making himself heard with some killer basslines. Dave Meegan is back as producer and he recreates some of the atmospheric texture I so loved on *Afraid of Sunlight*—h-era Marillion's finest hour in my opinion.

The disc opens with "**Between You & Me**," a straightforward rocker that'll knock you out if you're not careful. "**Quartz**" is up next. A total groove that plays two personalities against one another—the cold, unfeeling logical quartz watch vs. the delicate clockwork that needs care & maintenance. The song bumps & grinds with two outstanding Marillionesque moments—the atmospheric passage as h plaintively croons "it's so hard..." and the crashing silence that follows his visceral scream of "one of these days, you're just gonna ***stop***..." before the groove resumes. Play it loud. ☺

There's usually one song on every Marillion album which is simple enough (or sufficiently non-threatening) that I wonder why adult-oriented rock stations won't pick up on it. "**Map of the World**" fits the bill from *Anoraknophobia*. It's a beautiful little number about a girl with dreams of traveling the world, with lilting vocals and some great guitar work toward the end.

"**When I Meet God**" is a contemplative piece that has some serious depth. Like a fine wine, this one only gets better with time. It poses the familiar philosophical question of why a caring God would let bad things happen to its creations. The lyrics get a bit abstract at the end, so I leave any interpretations to you, the listener.

"**Fruit of the Wild Rose**" is another funky number that leads into another all-out rocker in "**Separated Out**." Samples from the movie "Freaks" are a bit overused for my taste but do make a good addition to a song that asks "am I enough of a freak, to be worth paying to see."

"**This is the 21st Century**" is a wonderfully moody composition. A drum machine sample serves as the backbone of the song, but layers of sound are built around and on top of it, creating a palpable texture that is surprisingly rewarding considering the slow tempo of the song. This is indeed the 21st Century and Marillion are still on top of their game.

Anoraknophobia closes with "**If My Heart Were a Ball It Would Roll Uphill**." It crashes all over the place like a monster and has a fantastic nightmare sequence that you can't wake up from.

In summary, once again they have created a totally original sound for themselves, but the indescribable essence of what makes Marillion "*Marillion*" is still in evidence. I applaud them for challenging their fans to journey with them to previously unexplored territories. It never ceases to amaze me how the first listen to their music leaves me thinking, "What the hell is *this*?" Yet given the proper time and attention, I can sink deeper into this music until I'm blissfully saturated in it.

The only thing left to say is: "Hey, Marillion! When can I pre-order the next CD?"

Today! I saw music in the sky, it sang around me I went blind
Like a masterpiece in a disguise, couldn't stop it pulling at my eyes
And I sang that pretty tune, an open airy song
And my heart looped the loop, well I could do no wrong.

- Lyrics by Steve Hogarth

Story 13

Let me tell you a little something about myself growing up. When it came to my search for that elusive female life companion, I sucked at it. I wasn't just the *Friend Zone's* most frequent visitor. I was the architect, the city planner and the mayor of the damn place. Over and over again, these wonderful women would come to know me, to care deeply about me, and yet not be interested at *all* in being anything more than friends with me. It's amazing how often I would play out that same scenario, never learning the proper lessons to instigate a different outcome. So like Sisyphus, I'd keep rolling that boulder up that same damn hill just to have it roll back down again, accomplishing nothing.

This next story is simply a musing on that longing I felt so often, as I would get *so close*, yet never quite cross that finish line. I went to college with a petite, intelligent, adorable, attractive girl named Jill. I've always been drawn to the smart ones—because who wouldn't? And I've always liked short women. Not just because I can't dunk, myself. But because, you know, they cost less to feed.

Also, I should note (to avoid any misunderstandings) that at this point in my life, Julie was (of course) a member of that tight-knit group of my closest lifelong friends, but not more than that... *yet*. It wasn't until years later, when she moved home from her job at Yale, that we started that years-long morphing of our relationship from close friends, to *best* friends, through that undefined gray area... to finally where we stand today—with her as the unquestionable partner-for-life I'm so happy to have.

Both at the time this narrative occurred and at the time I wrote this piece, Julie and I were still just friends. So rest assured this is not a betrayal in any way of the most important person in my life. I shared it with her at the time it was written. And as I look back on it now, I say what could be wrong with a little contemplation of young love and thinking, *Damn. To live those moments again...* Because when we think of young love, it's not really about the *object* of that love. It's about the *feeling* itself. That all-consuming, life-affirming, brain-scrambling, cold-sweat-breaking, stammer-inducing feeling of just being completely nuts about someone. And what a feeling it was. Over and over again.

I always seem to get the most positive feedback when I write about the opposite sex. So, either I write more passionately about passion, or you're all slaves to your hormones. I vote for the latter of the two. Regardless, I've searched my memory banks and chosen another romance-related episode to deliver. What it lacks in plot it makes up for in feeling.

Age 22, Graduate School, 1989
"Young Love"

She was so gorgeous with her honey-blonde hair and her radiant blue eyes and her petite figure and we got along just great but *dammit!*... She was going out with someone. And that someone was kind of a jerk to her. And that kind of a jerk happened to be a friend of mine.

Her boyfriend was one of those guys that for some bizarre reason women flock to because he doesn't treat them as well as he should. He often boasted to me (and probably all his other male friends as well) about his sexual conquests, regardless of the fact that that meant he was cheating.

I wanted *so badly* to tell Jill everything. I selfishly thought, "Dump this guy; he isn't worth it. Go out with me instead." But it wasn't my place. So anyway, enough background...

I had four tickets to a Jerry Garcia concert and it was going to be a great time. (Please don't think I'm one of those "Dead" worshippers. They're just another band.) I had a last-minute cancellation and Jill was the only one around who could go. So along she came.

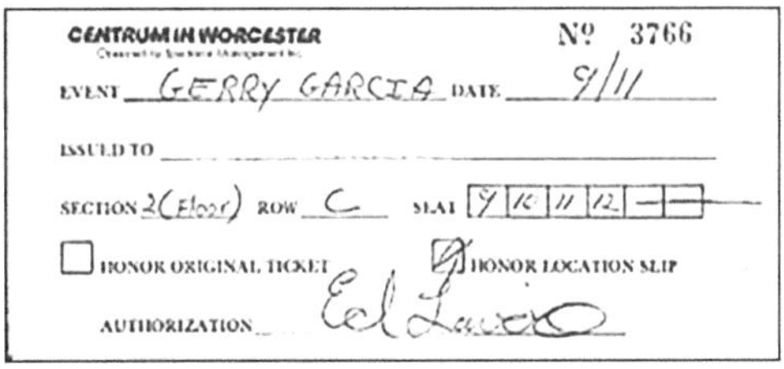
CENTRUM IN WORCESTER Nº 3766
EVENT GERRY GARCIA DATE 9/11
ISSUED TO
SECTION 2(Floor) ROW C SEAT 9 10 11 12
HONOR ORIGINAL TICKET HONOR LOCATION SLIP
AUTHORIZATION

I know. The ticket guy spelled Jerry's name wrong!

The concert was a blast. The music was surprisingly good. The atmosphere was joyous. The crowd was friendlier and more fun-loving than any I'd seen. And Jill was having as good a time with me as I was having with her. We sang, we danced (kind of), we laughed, we drank...

So the night had come to a close and we were back on campus. Jill had left her keys at my apartment, so we went up to get them. We walked in and I grabbed the keys off the kitchen table. I turned around and offered to walk her home. She lived right around the corner but it was late and Lowell *is* a city and OK, I admit it ... I didn't want the night to be over yet.

Right then, my roommate, Michael, came out of his bedroom to ask how the concert was. All of a sudden he apologized profusely and quickly retreated back to his room. (I later asked why he'd acted like he had just caught us making out or something and he said, "Didn't you see the look on her face? She wanted you to kiss her!")

So I walked her home and on the way, we laughed some more and talked about what a great time we'd had. A few minutes later we arrived at her place, much too soon in my opinion. She climbed the two steps to her door then turned and looked down at me.

This time I saw that look that Mike would later describe. She was beaming down at me. Her face was aglow and her eyes were sparkling. She said again what a wonderful time she had with me and... *she wasn't opening her door.* She was just standing there smiling at me.

This was it. This was my chance. It might never present itself again. One image flashed in my mind. I would climb those same two steps and sweep her into the most passionate kiss any lovestruck fool has ever given her .... A deep, tender kiss. The kind that sends shivers shooting up and down your spine.

Steve Martin said it quite eloquently in "L.A. Story": "Forget everything else and remember this: A kiss may not be the truth, but it's what we *wish* were true."

But alas, I didn't do it. Some hideous trick of nature has permanently embedded my psyche with a cruel set of moral standards known as a conscience. She was spoken for, and he was a friend (regardless of how good a friend he was).

Given another chance, forget morals, forget conscience, and forget a friend who I never see anymore in the present day. And today, I'd be remembering what I wished was true at the time. I love my life, but doesn't it seem there are way too many squandered opportunities when you think about it?

Story 14

When we think back on our childhood, I'm guessing those of us who celebrate Christmas would have memories from that time of year right up there at the top of our *Most Treasured Memories* list. Well, that surely applies to me. And it didn't hurt that I wasn't limited to only one "Christmas" each year. No. From the age of 5 until 13 (from divorce to Mom moving away), we had *multiple* Christmases. Pretty sweet deal for a kid with visions of presents (not sugar plums) dancing in his head each year…

Age 5, Grade 1, 1972
"Memories of Christmas"

The eventual happened when I was five years old. My parents got a divorce. Andre and Eleanor had come to a civilized separation of estate (as far as the word civilized can be stretched). Mom was keeping the kids while Dad had weekend visitation rights and found himself a nice little house less than half a mile away, of all places.

I was too young to remember the fighting mostly, so I didn't seem to think it was a big deal at the time. I just accepted the separation as another stage of life. And the fact that Dad was only a ten-minute walk away just served to reinforce that opinion. In fact, if anything, it made life better in some ways. A perfect example of one of those life improvements would be that wonderful time of year called Christmas.

Mom's parents would drive from Schenectady, New York, to spend Christmas Eve and Day with their now-single daughter and their two grandsons. They were old and slow while we were young and impatient, but they always came bearing gifts. So we loved them to death.

As a rule, we were allowed to each open one present on Christmas Eve before we went to bed. This year the one they chose was a mutual gift to the both of us. We started tearing at the wrapping of this package that was taller than me though skinnier.

"All right!" Alan exclaimed when enough of the surprise was revealed. "An air-hockey table!"

"All right!" I mimicked. "What's air hockey?" I asked, picturing the hockey players flying through the air instead of skating on ice.

"It's great! Let's set it up and we'll play."

Unlike the rest of the year, my brother and I actually got along at Christmas time. It was a happy time with all the presents and the lights and the rituals, and we stopped fighting for fear of fewer gifts from Santa. We were at each other's throats the rest of the year but we weren't *that* stupid come Christmas season.

First thing Christmas morning it was still too early to wake everyone up, but I couldn't stay in bed and pretend to sleep another minute longer. I snuck down the stairs in my airplane PJs and my slippers to find even more presents than were there the night before. Slippers were Grandma's *Big Thing*, as much as "Hats and Gloves" in the wintertime was Mom's rule.

"Put on your slippers or you'll catch cold!" Grandma would always say. And you didn't "catch *a* cold" with Grandma, you'd always "*catch cold*."

We would spend the morning in wonder and delight. Christmas was never a disappointment to the young. And we knew that even after all the presents here had been opened, there were still more to be had.

As the morning turned to afternoon, the two boys put a fresh set of clothes in a knapsack and hiked up to Dad's house. Here we got to live the Christmas experience all over again. And here we learned that sometimes the best things did come in little packages.

Dad watched his sons tear through all the presents knowing they hadn't yet opened the best one of all. Alan saw it first. "What's this?" he asked, as he held up a little ring-sized box. It had a tag on it that said, "Merry Christmas, Alan and Jeffrey. Love, Dad."

All the other presents had been from Santa, but Dad was taking full credit for this one.

"It's my present for you two," said Dad. "Merry Christmas, boys."

Alan was uncertain what to feel about this development. If this little thing was all he was getting from his Dad, was this something to start worrying about? *And I'm gonna have to share it with Jeff!*, was probably foremost in his mind at that point.

"Open it," I said, full of urgency. Maybe it's a gold nugget worth half a million bucks! "How heavy is it?"

"It doesn't weigh a thing," Alan answered.

"*Open* it!" I prodded again. Maybe it's a magic ring like the one in "The Hobbit."

Alan started to tear off the wrapping paper. I wanted to help with the opening process but knew that four hands couldn't find purchase on the little box, so I just sat there fidgeting. Alan lifted the top off as we both leaned closer in anticipation.

"It's a piece of *paper*!" Alan said in exasperation.

A piece of *paper*?

"Read it," our father prompted. Alan unfolded the small memo and announced, "It says, LOOK IN THE BASEMENT."

Time stopped for a moment as Alan and I locked eyes and held a silent conversation.

Something really neat must be down there!

Yeah! Whaddya think it is?

I don't know. That's why the garage door was locked and we had to come in through the front door instead.

Let's GO!!!

Dad saw his two boys freeze as the wheels turned inside their little heads. Then something clicked and (as one) they both rose and bolted for the stairway to the basement with cat-like speed.

"Careful! Careful!" he uselessly called. We couldn't hear him over our own screaming.

We stampeded down the dark stairs. Alan was a step ahead of me and got to flick on the light switch illuminating the basement. I stopped short and stared at the huge construction before me.

"A Ping-Pong table!" Alan yelled. "All right, *DAD*!"

Christmas evening was always spent at Dad's mother's house. She had taken her job of "going forth and multiplying" seriously, thus we had a slew of aunts and uncles. They would gather here to consume various confections, including pecan and pumpkin pies, cakes, cookies, the brownies with the chocolate chips and the walnuts (my favorite) and more.

Most importantly though to the younger kids, each pair of aunts and uncles could be counted on for at least one more gift. Usually, they were the boring kinds of presents (like shirts and sweaters and junk), but at least there were always a couple more good fun ones, like a good board game or something edible. As we grew older, a new rule was instated. Gifts for Godchildren only, not *all* the nephews and nieces. But these were the "good old days." We had *three* Christmases for a while there.

In the same spirit as the "Presents for Godchildren Only" tradition, every year the umpteen aunts and uncles would agree not to give each other presents, but every year a number of them would show up with presents for all anyway. The few caught off guard would protest, "But I thought we *agreed*! This year no presents for the siblings! I thought we really *meant* it this time!"

Christmas after Christmas I'd observe this same exchange. And every year I'd think the same thing, "What a bunch of dummies!" It was a mystery to me how these relatives of mine could actually be surprised when every year the same thing happened.

You see, I had figured out what I thought the rest of them didn't notice. Every year the people that showed up with the *unexpected* gifts were among those that were caught off guard the previous year. They must enjoy the silly charade somehow though, or else they surely would have stopped it years ago, right? Wrong.

Merry Christmas to All...

At this point, I want to clarify something. My brother is getting a pretty bad rap in these pages. And even though he'd be the first to tell you that *Yes*, he was indeed a horrible brother growing up, I have this to add as a counterpoint to that assessment. Something changed in the years between when I left for college and when I came home for that year of unemployment before I was able to start my career and move out on my own. He actually changed.

I know what you're thinking. People don't really change. We've all heard the tale of the frog and the scorpion. The scorpion can't help stinging that frog even though it means they'll both drown. Why is that? Because he's a scorpion. It's his *nature*.

Well, my brother is the only person I've known *in my entire life* who completely changed his nature. For the first 20 years of my life, he was the worst person I knew—this is not hyperbole. And somehow, some way, he transformed himself into one of the nicest people I now know. I can't explain it. I don't know how he did it, but it's not an act. He's now been this better version of himself for a decade longer than he was his lesser, angrier self. So the scales have tipped to nicety. And as the years go on, his generosity only increases.

Good for you, bro. Well done.

I guess there were the occasional smiles together in our youth...

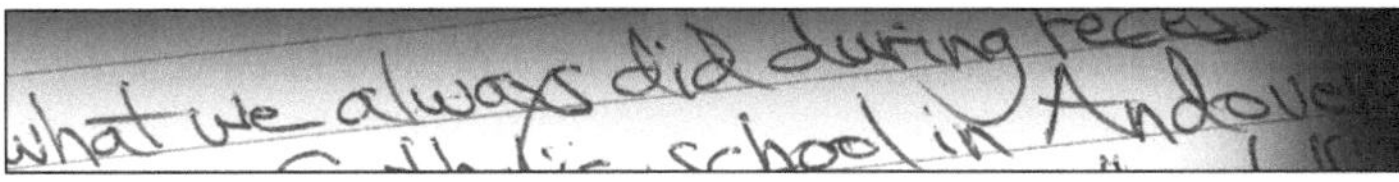

Story 15

I'm torn about this next piece. I like remembering how just the *thought* of a simple kiss could send an entire class of kids into such a frenzy. There's something beautiful about the innocence of that, and something bittersweet in the taking of that first step towards losing that innocence. Then, on the flip side, in a post-MeToo world I cringe at my assumptions of "being the man" and what I thought that meant at the time. It was a different world only five years ago. And this was written *twenty*-five years ago… about something that happened over *forty* years ago.

I've decided not to discard this one. Even though this is a tale of nothing happening, (the *nothing* being the attempt to attain a mere peck on the lips), I use it personally now as a reminder that I can be better not just in my actions, but in my thinking as well. Forgive me my stupidity. That's becoming a recurring theme here, isn't it?

Once again, I'll make the usual disclaimer about memory. I write this as I honestly remember it. Though I'm sure there's a lot of distortion in that.

Age 12, Grade 6, 1979
"The Club"

It was an important morning for all the kids in Grade 6 at West Elementary School in Andover. After school, the day before, Tim, Andy and I (slaves to our hormones and the onset of puberty) had decided that we were so obsessed with girls that we were going to start a special club. Initiation into the club required you to … *(Oh my God)* actually *kiss* a girl—*on school grounds*.

This caused quite a stir with all the kids in our grade. First of all, we had just started pairing off and few people had even kissed, much less kissed *here*, where the teachers could catch you. And most importantly, I was seeing Kim Sullivan and Andy was going out with Kim Beaton, but Tim… Tim wasn't seeing *anyone*.

Everywhere you went (in class, in the hallways…) the school was abuzz with the news. I don't know how the word spread so fast and so completely, but everyone was talking about it.

"Did you hear?…"

"Here at *school*?…"

"What's *Tim* going to do?…"

"He better not try to kiss *me*!"

At first recess break, Andy and Kim B. kissed. Most everybody hadn't realized he was going to make his move so quickly, so only the lucky ones got to witness the momentous occasion.

By the time second recess was upon us, Tim still didn't have a plan. So, I said that I would kiss Kim S. when class was over. The bell rang and everyone ran for the playground, except for Tim, Andy and me. We stayed back a minute to plan my strategy.

"Should I just walk up to her and do it?"

"What if she doesn't want to?"

"Don't worry, man. Of course she'll want to. You guys are going out, right?"

I walked out into the sunshine and fresh air with Tim and Andy in tow. Usually, Tim or Andy led the way, but not this time. I was the man, and I had a *purpose*.

To my surprise, the *entire* grade (boys and girls) were all waiting. They started cheering as soon as they saw me and I strutted my four-foot frame toward the crowd. As I approached the wall of eager kids, a gauntlet formed. They created an opening for me to walk through—a corridor of passage. The cheering grew louder and louder.

Now, I had never really been the most popular guy at school. I was kind of an enigma because I *hung around* with the popular kids even though I always got straight A's (OK, OK... A's *and* B's). It's not that we had already split up into social cliques. It's just that of the chemical mass of sixth-grade boys, I was allowed to orbit somewhat near the nucleus. Call me a first shell electron.

But not today... Today I was the one *everyone* was watching. Everyone was on my side and they were letting me know. I felt the power of the crowd surge through me, boosting my male ego like never before or never again. I was nervous earlier, but not now. Now, I was high on this seething throng of excitement and anticipation. With every step, I grew more and more confident.

And the crowd led me right to Kim. She was huddling with the other Kim and two other friends. I didn't stop. I couldn't. I had *power* and I could *feel* it. This was *great*!

I walked right up to her and swept her into my arms like I was Don Juan or Casanova. The crowd let out a roar like I had just hit the winning home run for the Sox in the World Series. Oh man, I was in my *glory*!

Then I looked at her face and what did I see... passion? longing?... uncontrollable desire for my sheer masculinity? None of the above.

Her face was a portrait of sheer mortification.

"P-please, Jeff. Not here in front of everybody," she pleaded.

I released her from my arms and wondered for a split moment if Andover Grade 6 would put up with such a letdown. Would they tear us limb from limb for having the audacity to not give them what they wanted? I announced to the crowd that we were going to do it later when they weren't watching (which we later did).

Never again did I draw so much attention from so many. But for a brief exhilarating moment, I was the focus of everybody's attention... and this time it wasn't even because they were laughing at me.

Chapter 25

An Un-Conventional Band

In 2002, Marillion held the first of their "Marillion Weekends" at Brean Sands Holiday Park on the Bristol Channel in the southwest of England. This convention consisted of three concerts on Friday night, Saturday night and Sunday afternoon. In 2003 and in 2005 they repeated their Marillion Weekends a bit further southwest on the channel at the Butlins resort in Minehead, England (now with a full concert on the Sunday night). In 2007 they moved to Center Parcs Port Zélande in the Netherlands.

Like many a diehard Marillion fan, Julie and I attended all four of these conventions. Traveling to Europe for Marillion became a near-biennial event for us, starting with that *Radiation* tour in 1998 (the one that ended with us at the MTV Europe Music Awards). We met some fantastic friends during those trips, including Rick Collins, who volunteered to drive us from London to the first Minehead convention without ever having met us. How great are Marillion fans?

The Rick Collins caravan on the way to (or from) Minehead. Hi guys!

The weekends continued to take place every other year through to the present day (as of this writing). From 2009 on, they did multiple weekends in multiple locations. Since it was much more economical for us, from that point on we attended all further conventions held at L'Olympia Theatre in Montreal, Canada. It was nice not to have to spring for a trip across the Atlantic each time. A five-hour drive north from

Massachusetts was a lot cheaper and easier than an international flight, not to mention the travel weariness.

These weekends were not just a couple of concerts strung together. Each convention featured one night in which a full album was featured. The other nights featured thematic concerts in which (in various years) they:

- Played songs from *A* to *Z*.
- Picked songs by pulling numbers from a bin for *Setlist Bingo*.
- Played songs from every year in descending order for *Tumbling Down the Years*.
- Let fans vote for their favorite songs for *Popular Music* night.
- Featured all their lesser-known songs for *Rarities and B-sides* night.
- Played just their longest songs and trilogies for a night of *Size Matters*.
- Spent one night playing all their singles.

In addition to the concerts, fans were treated to a weekend of Marillion-themed activities while occupying an entire resort filled solely with fans of the band. These activities included the following:

- Q&A sessions with the band
- Pub Trivia quizzes with Marillion-themed questions
- After-show parties
- A Marillion museum filled with artwork and artifacts—including the "mirrored overcoat" that h wore during the *Brave* tour. At a key moment during those performances, he would whip open this long coat exposing many pieces of reflective glass secured to the inside lining. A spotlight shining directly on him resulted in many frantic beams of light reflecting in various directions as if they were just exploding out of him. You could try on this coat during the "exhibit," along with many other items a Marillion fan would find compelling and intriguing…
- And then the most impressive bonus activity of all—"Swap The Band."

For this incredibly unique activity called "Swap the Band," all convention attendees were invited to send in recordings of their musical talents. Then, for an afternoon session in front of the full audience, several fans were invited to actually come up on stage and replace one member of the band and play with (or sing with) their favorite band *in concert.* How fantastic is that? Pretty damn fantastic if you ask me.

It was literally a dream come true for those lucky few who were selected. And for the audience, it was very much like a whole additional concert, as we got to hear the band playing a song with each Swap the Band participant. I also have to add that it was so wonderful to hear the entire crowd rooting and cheering for these talented fans with the same enthusiasm as they showed during the nightly shows. With some of them (especially the singers), you could see their incredible nervousness as they got started. But as soon as they started to belt out a tune, the crowd would cheer them on, boosting their confidence and thus lifting them up beyond that initial hesitance. This truly is a family. And family takes care of each other…

When Marillion announced their first convention, they said they were going to play the *Brave* album in its entirety for the first time since the *Brave* tour in 1994 (a tour I didn't get to see). I was so thrilled I remember getting up at like 4 AM to buy tickets directly from Marillion on UK time as soon as they went on sale. There was no way I was going to miss this convention.

What would be an amazing performance of *Brave* that first night of their very first weekend conventions was captured on video and released on the *Brave Live 2002* DVD.

In my typical purely pedantic, annoying fashion I complained online incessantly about the fact that the DVD was originally supposed to be the full Friday night concert. Yet it didn't include the 3 uplifting, emotion-releasing, high-energy encores performed that night ("Between You & Me," "Cover My Eyes" and "The Answering Machine.") Thus it was definitely **not** the full Friday night performance.

Oh, man. How did Lucy put up with us? Those encores showed up as bonus tracks on a later convention DVD if I remember correctly. So it was all unnecessary whining on my part.

These multi-night convention weekends may seem like an original idea. Yet they aren't listed among Marillion's "firsts" from a few chapters ago. That's because the inspiration for these conventions came from elsewhere. In the Band Interviews on the *Brave Live 2002* DVD, Mark Kelly tells us they actually got the idea from the band, The Stranglers. In 2001, The Stranglers held their first of five 2-day conventions between then and 2011.

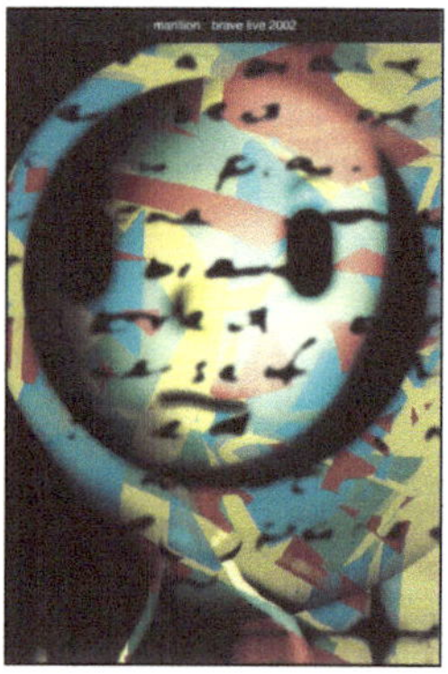

Marillion took a great idea and made it their own—expanding it to a full 3-day affair. These Marillion conventions started in 2002 and continue to this day. Their planned 2021 convention was unsurprisingly delayed due to the Coronavirus. Yet they hosted a three-night virtual convention in September of 2020 at no charge, called the *Couch Convention 2020.* A full concert film was streaming each night on their official YouTube channel. The band was watching along with the fans so we all experienced the shows "as one." Additionally, there was the usual assortment of Marillion Weekend experiences with quizzes, Instagram chats, Q&A sessions and after-show "parties."

If you knew your Marillion trivia, you could be led on a Marillion *Scavenger Hunt* on their website. Clues and riddles led you to a series of couch icons. Clicking on each icon provided you with a character (or letter), as well as the clue to the next couch icon location. All these characters, when strung together in the correct order, spelled out the address of a website where the *Golden Couch* resided. I didn't win, so I can't tell you what the prize was. But it was a fun exercise during lockdown.

Forgive the poor quality of this picture (right). After completing the challenge, I snapped a picture with my phone of my computer screen in case the contest entry info disappeared seconds later, Mission Impossible-style.

Age 34, 2002
The First Marillion Weekend

As the date for that first Marillion weekend approached, I was so ecstatic I wrote up a ridiculously long email to send to all my friends. Seriously, it must have been like 10 email pages. I provided a song-by-song analysis of the *Brave* album to show them what they had been missing out on all these years. I hit *Send* and, of course… most of them probably never even read it. So here in its entirety is the evidence of my folly and the next installment of our Rocking Out segments. This isn't a review; it's technically more of a *pre*-review. Let's just call it a pre-view. I'm tempted to correct or update a few things but I'll leave it as is since at the time this is what these songs meant to me.

I get a real laugh seeing this today and remembering how I honestly thought (after all the years that came before), that even one person in my life would be moved by my words enough to listen to Marillion's *Brave*. And how insane was I to spend that amount of time crafting and constructing this absurdity of letters? Ha. *Great, another email about Marillion from Jeff… Jeez, how frigging* **long** *is this thing?… Delete.*

So, riddle me this, Batman. Since I'm going to give a copy of this book to most of those friends. And since that ludicrous diatribe is now *in the book…* Does that mean they're now going to actually read it? Vengeance is *mine*!

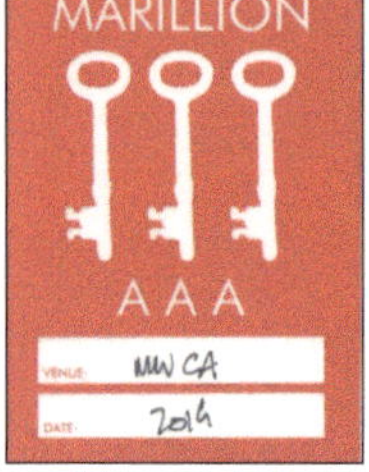

----- Original Message -----
From: Jeffrey Pelletier
Bcc: Everyone
Sent: Saturday, March 23, 2002 12:14 PM
Subject: BRAVE

Okay. A week from now we'll be heading to England once again to see my favorite band Marillion – though this time it's different. After several days hanging out in London, we're heading down to the coast to a seaside resort in Somerset for the first worldwide Marillion convention.

We have a full concert Friday and Saturday night as well as a lunchtime acoustic show on Sunday – plus all sorts of Marillion-related activities and several great opening bands (including The Wishing Tree – Steve Rothery's solo project with a wonderful female vocalist). The resort holds 2000 people and it is completely sold out exclusively by Marillion fans with more that are wishing they could come. People are coming from all 4 quadrants of the globe – Japan, Australia, South Africa, Argentina, Europe, North America...

But the thing that has me more excited than anything else is this... For the first time since it was first recorded and toured in 1994 (a tour that didn't come to the states), Marillion will be playing their dark, masterpiece of a concept album BRAVE from start to finish. For those of you who don't care, stop reading now. I won't be offended or disappointed. For those of you who *are* interested, here's the scoop on Brave...

Before Steve Hogarth ("h") joined the band in 1988-89, he heard a police broadcast on the radio. A traumatized girl had been found wandering on the huge Severn bridge (think "Golden Gate" in the UK). She had no memory of who she was and couldn't even speak. The police in a final effort appealed to anyone who might be listening, in an attempt to locate someone that knew her.

"h" keeps journals. And he wrote about this girl, wondering what had driven her to this state. Fast forward to 1992-93. Marillion are recording their latest album. The pressure is very high from the studio to come up with another hit (like "Kayleigh"). It's been a long time since they had a radio song that big and the record company is losing its patience. In a ballsy move, Marillion choose to go the exact opposite direction.

The few songs they've written remind h of the girl from the radio years ago and he pulls out his old journals. He proposes that they do an entire album about this girl. Fill in her painful past with stories of abuse and an inability to love or let herself be loved – the motivation that drove her to the bridge and what would happen to the fictionalized version of the girl once she got there.

Marillion spent the next year and a half orchestrating and fine-tuning the most remarkable piece of music I've ever heard. It's certainly their least accessible album. It's very dark and takes many listens before you even start to appreciate its depth. It was months before I could truly appreciate this album. I have literally listened to Brave hundreds upon hundreds of times over the years and if I allow myself to be taken by the story, it still brings tears to my eyes to this day. And I admit that shamelessly.

The following is a track-by-track dissertation on Brave with relevant sections of lyrics.

"Bridge" – the perfect mood-setter, the water laps against the shore and a distant foghorn grows louder with each blast merging into a plaintive synthesizer chord. If you listen closely, you hear the girl whisper "a bridge is not a high place."

"Staring out over the bridge, a million photo flashes from the water down below
Dawn light bouncing through the mist. Roar of traffic and the crack of police radios..."

"Living With the Big Lie" – we go all the way back to the moment of her birth ("The bright light") and are taken through her childhood. The pain of growing up and realizing that the world isn't as kind a place as you thought it was in the innocence of youth.

"It all began with the bright light, the bright light and the noise
The chaos of the senses and the scream of desire
Touching and being touched, a million loose ends to tie up
Ticking of the clock and the cradle rock

"One fine day, the chaos subsides
Bleeds into awareness and a lifetime of surprise
The beauty of your mother's eyes, the pain when you fall
You drink it in and marvel at it all, but you never really figure it out. You get used to it ...

"The attitude of authority, the laws and the rules
Hit me square in the face, the first morning at school
The heroes and the zeroes, The first love of my life
When to kiss and to kick and to keep your head down when they're choosing the sides
I was never any good at it, I was terrified most of the time
I never got over it... I got used to it..."

"Runaway" – the climax of the first part of the album for me. She's a teenager now and she's off on her own. The abuse has driven her onto the streets and the tension builds to an emotionally powerful musical crescendo that gives me chills each time I hear it…

"Were they deaf to the prayer behind your lies
Maybe they didn't care to ask you why
Was a runaway girl all they chose to see
All they were prepared to admit to…

"So you cower in the town's forgotten places
And make your bed with unfamiliar faces
And at last you've got your freedom but that's all you've got
You're trying to make your mind up if you're better off

"You pretend to wait for washing in some Laundromat
But you're damned if you'll give them the satisfaction of you going back
You'd freeze before you share a roof with them
And you'd starve before you let him get his hands on you again
Did you cry when they dragged you home… "

"Goodbye to All That" – An extended piece made up of several passages chronicling her descent into the seedy street life, the drug culture and being used for her body.

> **Wave** – *"Hold your breath… 'til you feel it begin... Here it comes…"*
>
> **Mad** – *"Was it something I didn't say, or was it something I said? Leave that stuff alone, use me instead…"*
>
> **The Opium Den** – *"Roll me up and breathe me in, come to my madness, my opium den, make sense of it again…"*
>
> **Slide** – *"Make a drug of my affection, I'll get you out of your head, Love wastes us together, love holds us together, love folds us together…"*
>
> **Standing in the Swing** – *"I'm ashes on the water now, somewhere far away, you think you came here just in time, but you're 20 years too late, you won't have long to wait…"*

"Hard as Love" – a powerhouse song detailing her inability to love, she's been hurt too much by it…

"Well I hear that you were looking out to take me home
And I hear that you can handle it and you're not scared
Have you heard about the pictures on my bedroom wall
Baby nothing else has ever been as hard as love

"It makes you humble and it makes you crawl, it makes you tremble and it makes you fall, but it's all worthwhile…

"Well I hear that you were looking out to change my faith
You can bash me with your book of words, hammer me down!
But you know that love can be as hard as six inch nails
Baby nothing else has ever been as hard as love…"

"The Hollow Man" – a tender ballad with several interpretations, the father (or stepfather) who feels empty because he knows the terrible things he's done or the boyfriend who couldn't reach her...

"I think I have become one of a pair of men
It's a feeling I've been having for some time
I look down upon myself and watch my movements
A blind eye sees the fragile vandalized...

"Watch this cold world dishing up these endless hollow men
Find us anywhere you look, come and meet our friends
Stand us in our silly clothes, put our batteries in
Line us up like fairground ducks, watch us grin and grin
See the lies behind our eyes, see the will to win
We'll buy you and we'll sell you, but perhaps we'll save your skin

"We're tin-hard and we rattle when we're shaken..."

"Alone Again in the Lap of Luxury" – a detailed look at the home life she must have come from. The mother who turns a blind eye and meticulously cleans and cleans the house as if that will justify what takes place inside it...

"He's been talking in his sleep again,
Sayin' he's sorry, calling my name,
Saying how he's so ashamed...

"Since it happened I had nothin' to say, It used to bother them but now it's OK
Mother cleans his dreadful house every day, Scrubbing at the stains that won't go away
They sent me away to the school in the park
They said it would be good for me
Still hear the other kids cry in the dark...

"I don't remember the last time I cried, I don't remember much except lies
See the little girl spirallin' down, this is a photograph of who she is now...
Alone again in the lap of the luxury...

"I could be anywhere right now
If I only had the nerve to leave this house
Maybe somewhere by the sea
Take me somewhere, anywhere please!..."

"Now Wash Your Hands" – another short mood-setting piece.

"You give up hope, you settle down, with your favorite soap. Now wash your hands..."

"Paper Lies" – Here we have one of the few Marillion songs I'm not crazy about – mostly because I think it's the one piece that doesn't fit into this concept – a topic much debated amongst the fans. It's about the media and their sensationalism and perversion of truth, any effort to make a story.

"When you look into the mirror do you see a face you hardly recognize
If you look into the sun too long, do the things you see begin to burn your eyes
When you kill the truth, you can make a killing
You might just make losing look like winning..."

"Brave" – We now begin the slow ascent to the bridge and the climax of the story. Uilleann pipes lead into an atmospheric musical setting where a stunning vocal extols our Brave girl...

"What a brave brave girl, never lied before
Such a plain deceit, everyone would eventually know...

"What a brave brave girl, never loved before
Placed herself in reach, so he tried, in his own way,
To find the heart of the tight-packed rose

"She's gone now... She's gone now...
But oh... she aches.... Ohhh, she aches!........."

"The Great Escape" – In my opinion, the most emotional song ever written. A piece in three parts. The girl is poised up on the bridge. We now understand everything that's led her to this moment. She leans out, ready for the great escape, to jump...

"Heading for the great escape, heading for the rave, heading for the permanent holiday
Heading for the winter trip, heading for the slide, heading for the dignified walk-away
Heading for the open road, goodbye to all that, heading for the automatic overload

Standing in the open boat, standing in the swing, waiting for the ringing and the white light
Waiting to be recognized, quiet applause will do,
They shower you with flowers when they bury you...

"You're holding on... You're holding on...
You're hoooooooooooooooooooooooooooooooooooooooo....."

"The Great Escape: The Last of You" - She's done it, she jumped. There's no other interpretation for that plummeting scream we just heard. As she falls to the water below we have these final words from her, sung with a ferocity that grabs you by your heart and won't let go...

"Just when I thought I'd seen the last of you, you come here scratchin' at my door
Your pain and anger's in the howling dark, of every corridor I walk
So tell me more, about the love that you rejected
Tell me more, about the trust you disrespected
I still don't know, why did you hurt the very one
Why did you hurt the very one that you should have protected?......"

"The Great Escape: Falling From the Moon" – sung with a sad resignation...

"Don't ask me why I'm doing this, you wouldn't understand
You're asking the wrong questions, you couldn't understand
A bridge is not a high place, the 52nd floor, Icarus would know,
A mountain isn't far to fall, when you've fallen from the moon...

"There's murder on the street
I'm ashes on the water now
Somewhere far away
I have fallen... fallen... From the moon..."

"Made Again" – Now here it gets interesting. The original version of the album faded into an orchestral outro which comes to rest on the lapping of the water on the shore once again. But Marillion, the studio, the producers, everyone thought this was just too dark. So this light acoustic piece is what officially closes the final version of the album. Another hotly debated topic. Some accuse Marillion of selling out to the happy Hollywood ending. Others find it the perfect closing note.

It can have more than one interpretation. She's dead and her spirit has been made again in the religious sense. Or she survived the fall and the shock of it has made her again emotionally, so she can begin the healing process... Ask me, I like the song. It's not a happy ending. She's transcended onto the next plane so to speak (not that I believe in that). She's gone.

"I have been here many times before in a life I used to live
But I never saw these streets so fresh, washed with morning rain
Like the whole world, has been made again

"And it's all because you made me see, what is false and what is true
Like the inside and the outside of me, is being made again by you
Like a bright new morning, like a bright new day...
I woke up from a deep deep sleep, I woke up from a bad dream
To a bright new morning, to a bright new world...
Like the whole world has been made again..."

So there you have it – *Brave.* If asked, I would say that *Afraid of Sunlight* is my favorite album. However, *Brave* is their best. (A subtle distinction.)

On Friday, April 5th, Julie and I will be seeing this piece done in its entirety. I've seen them play "The Great Escape" three times before (once with the song "Brave" leading into it) and it's not uncommon to look around and see eyes filled with tears on the majority of the faces surrounding you in the audience. With the entire story preceding it this time, this is going to be the most powerful musical experience I anticipate I'll ever see and feel.

For anyone who would like to give *Brave* a chance with an open mind and a willingness to listen to it more than once, I'll gladly provide you with the music. It is such a well-produced CD as well. There are many things going on in the background that you will pick out with each subsequent listen – subtle whispering throughout, a mother and daughter screaming at each other in "Runaway," a baby crying in "Brave," a splash at the end of "Paper Lies," the lapping water, the foghorn...

To anyone that actually read all of this. Thanks for your time. If just one of you actually did read this and has an appreciation of what I'm so excited about, then it was worth the hour it took to write it.

Take care!

Jeff P

Between sets at the Minehead convention (left) and enjoying the pub quiz at Port Zélande (right).

Chapter 26

"So Many Levels…"

All right. How many of you actually read that *Brave* dissertation? I mean, really *read* the whole thing. I'll be shocked if even half of you did. But that's OK. It's your time. I said at the beginning you can do with this book as you see fit. And you don't need that deep of an understanding of the album to appreciate the following story. You get that *Brave* was a complex musical journey.

So let's jump back in time for a moment. We're back at the Somerville Theatre during the Tour Fund tour in 1997. After the show, a group of us had formed a circle around h in the parking lot, as he generously signed whatever we were handing him. At this point a well-intentioned fan turned to Julie and asked, loud enough for all of us (*including h*) to hear, "So… what do you think *Brave* is about?"

At this point, h stopped signing the imported *This Strange Engine* CD sleeve I had given him (see below) and looked up expectantly at Julie. Immediately, I thought, *Oh no. Please do* not *start trying to explain Brave to this guy in front of the man who actually wrote it…*

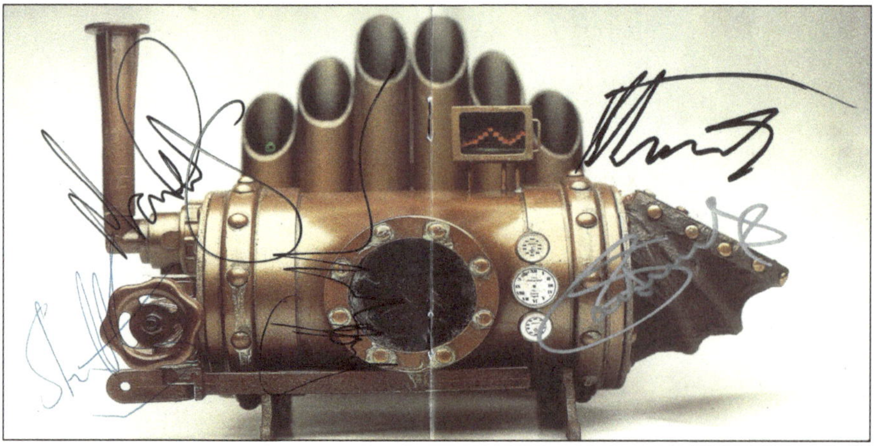

I never should have doubted her. With just enough of a dismissive tone to bring the conversation to an end, she said, "Oh. It's about *so many* things, on *so many* levels..."

H nodded matter-of-factly and uttered his agreement just loud enough for me to hear... "*Mm-hmm.*" He then turned back to his task at hand and resumed his autographing.

I have never been more proud of that girl.

Years later, h giving Julie a delighted hug in Montreal upon recognizing her as "Julie from Harvard."

Chapter 27

Pre-Order Redux

Remember that string of three Marillion albums in three years (1997, 1998 and 1999)? With the pre-order campaign a success, they didn't have the usual record company pressure and they were able to take as much time as they wanted on *Anoraknophobia*. It came out in 2001, two years after their previous album. After the usual amount of touring to promote the album, they then set to work on new material. Once again, they had the luxury to take as much time as they needed. How much time? A full *three* years, as it turned out. But there was a silver lining for the fans enduring this long wait. During those three years, they had written enough music that they deemed worthy to justify *Album 13* being a double album—a first for the band.

For their thirteenth album, they decided to repeat the successful pre-order campaign that they used on their twelfth. Retaining the musical rights to *Anoraknophobia* must have paid off as expected because now they had a different plan. Fans would once again be given the opportunity to pre-order a Marillion album and be able to see their names in the liner notes. However, this time there would be two options. They could pre-order a standard single CD or they could pre-order the special *Campaign Edition* double CD.

The funds they would collect for the *Campaign Edition* would go into a special marketing fund designed to raise awareness of the band. This would include both radio and magazine adverts targeted at the outlets they knew their fans liked, as well as one of their largest touring efforts to date. They even paid to have fans' cars wrapped in the album artwork (seen on following page). Additionally, when the album was to be released, for the first few months it would be on sale exclusively through their website. A distribution deal to sell it by standard means would follow several months later. This would maximize the portion of their album revenue that would go directly to the artists themselves.

Marbles promo card (left). And "wrapped" Marbles car (right) seen at the Butlins convention in Minehead. So cool that Mr. Saks had to have his picture taken with it. Who's the proud owner of that vehicle? Let me know and I'll give you a shout-out.

In July of 2003, it was announced that the 13th album would be called *Marbles* and the 2-CD Deluxe *Campaign Edition* of the album would be housed in a 128-page hardcover book package filled with pages upon pages of artwork, fan names, lyrics and liner notes. To this day, this is possibly both my favorite Marillion album and the most gorgeously packaged (rivaled in my mind only by the similarly impressive special edition of Kevin Gilbert's *The Shaming of the True*).

Once again, the world was starting to take notice of Marillion and their groundbreaking ways. The following article was printed in the Boston Herald to my delighted surprise. Note how at the bottom it repeats that phrase that started my live Marillion odyssey... "Marillion plays the Paradise in Boston." Though this time I was very much aware of the upcoming show at The Paradise Rock Club. And the following page contains the official Campaign Fund announcement from the band.

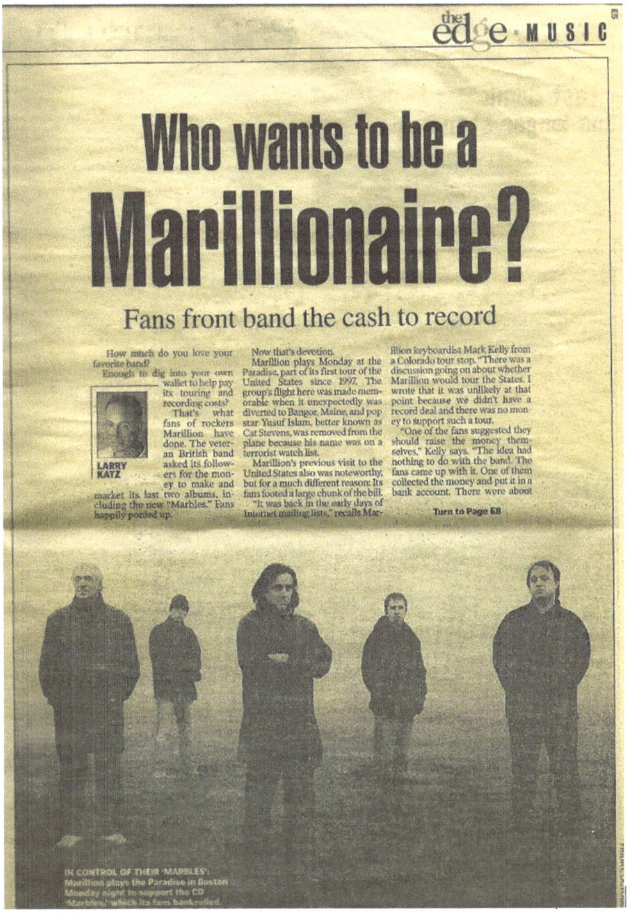

the edge · MUSIC

Who wants to be a Marillionaire?

Fans front band the cash to record

LARRY KATZ

How much do you love your favorite band?

Enough to dig into your own wallet to help pay its touring and recording costs?

That's what fans of rockers Marillion have done. The veteran British band asked its followers for the money to make and market its last two albums, including the new "Marbles." Fans happily ponied up.

Now that's devotion.

Marillion plays Monday at the Paradise, part of its first tour of the United States since 1997. The group's flight here was made memorable when it unexpectedly was diverted to Bangor, Maine, and pop star Yusuf Islam, better known as Cat Stevens, was removed from the plane because his name was on a terrorist watch list.

Marillion's previous visit to the United States also was noteworthy, but for a much different reason: Its fans footed a large chunk of the bill.

"It was back in the early days of Internet mailing lists," recalls Marillion keyboardist Mark Kelly from a Colorado tour stop. "There was a discussion going on about whether Marillion would tour the States. I wrote that it was unlikely at that point because we didn't have a record deal and there was no money to support such a tour.

"One of the fans suggested they should raise the money themselves," Kelly says. "The idea had nothing to do with the band. The fans came up with it. One of them collected the money and put it in a bank account. There were about

Turn to Page E8

IN CONTROL OF THEIR 'MARBLES': Marillion plays the Paradise in Boston Monday night to support the CD 'Marbles,' which its fans bankrolled.

THE NEW ALBUM CAMPAIGN

After much deliberation, we have finally decided on a name for the new Marillion album - it's going to be called Marbles. This album will also be a Marillion first - a double-album! Yes, that's right, a full 2CD's of new music.

Since we first announced our initial ideas about releasing the album over eight months ago, we have been flooded with emails full of your comments and suggestions. With help from your feedback, we have now formulated the final plan! (You may want to sit down with a cup of coffee - there is a lot to read!)

Shortly, we will be putting the new album on sale for pre-order from our website. For every copy of Marbles that is pre-ordered, profits from that sale will go into our Campaign Fund. Using this money, we will take out adverts in magazines that we know our fans read, advertise on radio stations that we know our fans listen to - in short, try to reach all the people who don't necessarily visit our website or receive our eWeb email newsletter, and potentially reach NEW fans too!

Radio advertising is a very important medium for us, as we are sure you will agree - if people can hear our music then they generally love it. As radio stations will not play Marillion songs (politics and playlist restrictions - DJ's have told us they're not ALLOWED to play our new music!), then this seems an obvious route for us. The Campaign Fund will allow us to take control of our marketing and destiny.

As of 1st September 2003 Marbles will be on sale - only through Racket Records and www.marillion.com.

The Marbles Deluxe Campaign Edition (in very special packaging!) will be on sale through Racket Records at the special pre-order price of £27.99. If you purchase the Deluxe Campaign Edition before 31st December 2003 you will be directly helping our Campaign Fund, and as a "Thank You" will we print your name in the album credits, AND ship the album to you at least a month before the currently planned release date of 26th April 2004*. Remember, this is a double album - 2 full CDs of new Marillion music!

As of 1st January 2004 onwards you will still be able to buy the full double album, but your name will not be in the credits, and it will not be shipped before the release date; but it will still be the Deluxe Campaign Edition packaging. For those of you who want all the music but don't want to pre-order and take part in the Marbles Campaign, we plan to release the full double album in a standard CD box (price to be decided) from the planned release date of 26th April 2004*, available from Racket Records.

The current planned release date for Marbles is 26th April 2004. From this date, there will be a 1 CD version of the album for sale in retail shops around the world (the full 2 CD double album will NOT be available in retail shops, only from Racket Records). When this 1 CD version has gone on sale in the shops, we will also be selling it via our website.

A major world tour will begin at the end of April 2004, and we will hopefully start announcing dates early this autumn.

We will also be selling special Marbles Campaign Merchandise, which will also help raise money towards our goal. This will be on sale from 1st September 2003 - our plans are for posters, t-shirts, gold discs, and other goodies! Everyone who orders ANY Campaign Merchandise will automatically be entered into a free prize draw. Prizes include soundcheck passes, personalised gold presentation discs, autographed CDs and DVDs, and some very special tickets to a pre-album listening party. In February 2004 (date to be confirmed), we will be holding listening parties around the world to allow fans and journalists the chance to hear the brand new album. These will be held in the USA, Canada, France, Germany, Holland, UK, Italy, Spain, Norway and Poland; and each one will be attended by a member of the band.

Every time you buy a piece of Campaign Merchandise we will enter you into a draw to win a one of these great prizes, or even the Grand Prize: The Super Ticket. This is the prize of a lifetime for one Marillion fan: two concert tickets and backstage passes to any Marillion gig on the forthcoming 2004 tour. Not only that, but we will also pay travel and hotel expenses for the lucky winners!

As for Marbles, we don't have a full track listing yet, but sound clips and album updates will be available at www.marillion.com in the near future. One thing is for sure: it WILL be 2 full CDs of new music - it is a double album without a doubt!

Marbles is still taking shape, but the few outside the band who have heard it are buzzing. Thanks again to all of you for your faith and continued support.

H, Mark, Ian, Steve, and Pete

For more details on the Marbles Campaign, Campaign Merchandise, and regular updates, visit: www.marillion.com/marbles

* At time of going to press, the scheduled release date for 'Marbles' is Monday 26 April 2004. Please visit www.marillion.com for any updates or changes to this date or other pre-order information.

Chapter 28

Cheating the System (*a little*)

The band felt confident in both the way they were now running their business and in the music that they had recently created. It had been quite some time since they had charted a hit single into the Top 40. Yet when they looked at actual singles sales, they realized that the market was much smaller than it used to be. A single that sold only a tenth of what "Kayleigh" sold would probably be a Top 10 hit now. Once again, they came up with a plan.

"You're Gone" would be the lead single from *Marbles* and would be released in April of 2004. There would be three different versions of the single. Two versions would be released on CD with different bonus tracks. And a third version would be a DVD single. All three versions of the "You're Gone" single would be eligible for chart consideration as a cumulative total for that song. Sure, that's been done before. But here's where it gets interesting.

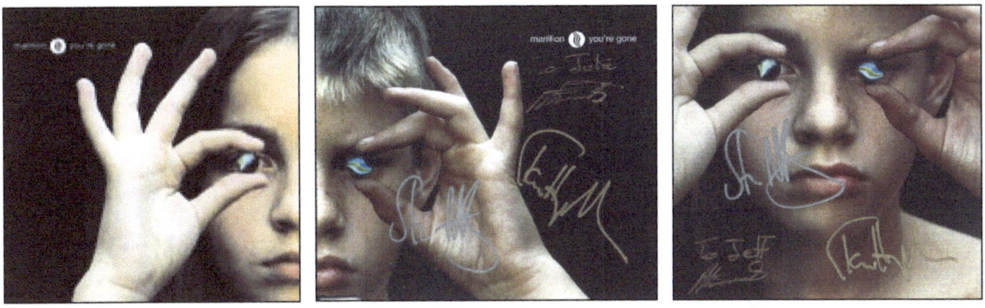

You're Gone—CD Single 1, CD Single 2 and DVD Single

Most of their fans around the world would need to import these singles as they often did, not being able to buy domestic releases in their home countries. Marillion suggested the following… We're a big family here. A community. Instead of importing these singles through Amazon (or some other online outlet), why not contact a local in the UK to buy them for you at their neighborhood record shop? They suggested that if a large enough portion of their international fans made their purchases chart-eligible by obtaining a local "Singles Buddy" to do the actual purchasing for them, then there was a very good chance that they would have a hit single—maybe even a Top 20 single.

You've seen this movie before now, haven't you? You know where this is going. Marillion fans coordinated their sales and made such a splash with their activity that the charting service warned that Marillion could not be in charge of organizing this effort or the single would be deemed ineligible. The band stood back and let what would happen play out on its own...

The week after the single's release, the band listened closely to the chart announcements. The bottom 20 songs in the Top 40 were read with no sign of Marillion. They collectively leaned forward as the next 10 names were read (#20 - #11). Again, no Marillion song. Was the plan a bust? Did they not even reach the Top 40? The final names (the Top 10) were then announced. When they reached number seven, *there it was...* "You're Gone" by Marillion at #7.

Marillion had their first Top 10 single in almost 20 years (since "Incommunicado" from 1987's *Clutching at Straws*). It also reached #8 in the Dutch Top 40. What followed was a resurgence in media interest in the band that most of them had thought had faded into the sunset long ago.

The band repeated the process with their second Marbles single, "Don't Hurt Yourself." In July of 2004, they released two versions of this second single and broke through for another entry into the Top 20. "Don't Hurt Yourself" reached #16 on the UK charts (and #35 in the Netherlands).

"Don't Hurt Yourself" – Blue CD Single 1 and Red CD Single 2. Again, procured for us by our "Singles Buddy," Rick Collins. Thanks, Rick!

And so, the band that hadn't had hit singles in many, many years found a way to make it happen by bending (but not *breaking*) the rules. Essentially, once again, Marillion was behind doing things in an original way, an *innovative* way. Marillion's fans are unlike those of almost every other band. Yet, on top of that, the band is fully aware of what it has in its fans and knows how to capitalize on that potential.

Chapter 29

Marbles Have Bro-o-ken (Like the First Mo-o-orn-ing)

The *Marbles* tour was the first to come to the United States in seven years. The Tour Fund was back in 1997, and 2004 had been a long, long wait between tours. On September 21st, the band and crew boarded a flight that was supposed to connect them in Washington D.C. through to Mexico City, the first North American date of the *Marbles* tour.

Instead, they were informed over the PA system that their flight was being grounded at Bangor International Airport in Maine. H distinctly recalls looking out his window and seeing several large military aircraft and thinking that it looked like a military base.

Over the PA they were then told that the unscheduled landing was due to bad weather in D.C. and the need to refuel. The feeling on the plane got more tense when they were then informed that they were not allowed to use their cell phones. After another delay, the voice over the PA told them that he was given permission to tell them that the diversion had been requested by the FBI and security forces. The voice said that he was "forbidden to fly to Washington in this aircraft with these passengers and this crew at this time." Holy crap. That did not sound good…

Later, they were all ordered to de-plane with all their belongings. A crewmember let it slip that two suits from the FBI had already removed two people from the aircraft. After a couple of hours of being detained in the international terminal while passengers were screened and interviewed, h found an airport official who actually answered his question. In a hushed tone, he conveyed, "The FBI just took Cat Stevens and his daughter off your flight."

H discreetly made a call to manager Lucy, updated her on the situation and spilled the news. With their first concert date now in jeopardy, Lucy said she was on it. They eventually re-boarded the plane and were flown to D.C. after a seven-hour delay. They had missed their connecting flight and would have to fly out in the morning. No hotel rooms at all were available due to the popularity of a new Native American history exhibit at the Smithsonian. And so they had to sleep in the terminal, under bright fluorescent lighting.

Can we just get something straight here? Cat Yusuf Stevens is a pacifist. He is now and always has been. He wrote songs of love and peace. He never once said he supported the fatwa against Salmon Rushdie. At a lecture in 1989, he apparently answered a question about blasphemy and Islamic Law and said that as far as he understood, the Qur'an considers it a capital offense. He never once said he agreed with that, nor called for anyone's death. What has happened to him because of the way that remark was misrepresented by the press, is (in my opinion) one of the most shameful and undeserved public shunnings in modern cultural history.

And I would like to add that Cat has still been producing beautiful music in this current century, in addition to the previous one. His song "Heaven/ Where True Love Goes" from 2006's An Other Cup *may be the best song he's ever written. Go listen to it the first chance you get. It will probably be the best four minutes and fifty seconds you spend all day. (Don't sell it short; play it loud.)*

After only an hour of sleep, though, h's phone began to ring. He then found himself being interviewed by BBC Radio 5 live, followed by Good Morning America and Channel 4 news. He was still on his phone up until the point he boarded his flight (now with further delays by connecting through Chicago). All of this while he was scheduled to sing in front of 9,000 people at the National Auditorium *that night...* By the time the band arrived in Chicago, they had been traveling for 33 hours straight and were still an entire country away from Mexico City.

The band did in fact make that show... just barely. Most of us remember when the story of Cat Stevens' immigration issues made headline news back in 2004. What no one realizes is that this was another major story that was tied to Marillion. The story was "broken" by Steve Hogarth.

For the full story in vivid detail (and many others), pick yourself up a copy of "The Invisible Man Diaries" by Steve Hogarth (*cover design by Robert Hammond*). This story is in the second volume, "1998 – 2014."

Chapter 30

Firing on All Cylinders

Julie and I are very fortunate that we could afford to travel for two UK tours and three European conventions in those seven years between American tours. We figured, however, that this may be our last chance to see Marillion in the States. We also were nuts about the new music. So we decided to go wild and ended up seeing four shows on the *Marbles* tour. We traveled to Chicago (where we were joined at the show by brother Mark Hanlon and nephew Pat Hanlon) and then to Philadelphia. The following weekend, we again saw our "local shows" in Boston and New Haven.

New Haven setlist (left) and Julie & h (right) at the Border's signing session in Boston.

How do I explain how amazing that tour was for us? How can I convey to a layman the feeling of traveling from night to night of incredible music, (as the Red Sox were surging in the playoffs, before mounting what would become the greatest comeback in playoff history against the century-long rival Yankees)?

From the ominous opening notes of the first song, "The Invisible Man" (where h would shuffle out to the microphone like Gene Wilder's *Willy Wonka* about to open the gates to his magical factory of chocolate and child endangerment), to the set-closing exultation of "Neverland" (where h would sing of the love that any fool can see inside of him, and then emulate soaring on the waves of the love from the crowd, and eventually spread his arms wide as if gathering up all of that love and affection as the crowd cheered louder and stronger, and then slowly closing, containing, embracing, preserving, squeezing tightly and absorbing it all inside him).

There was truly something magical happening on those stages and in those audiences. This will forever be imprinted in my memory as the most enjoyable tour I've ever experienced.

You provide the soul, the spark that drives me on
Makes me something more than flesh and bone
At times like these, any fool can see, any fool can see, your love inside me

- Lyrics by Steve Hogarth

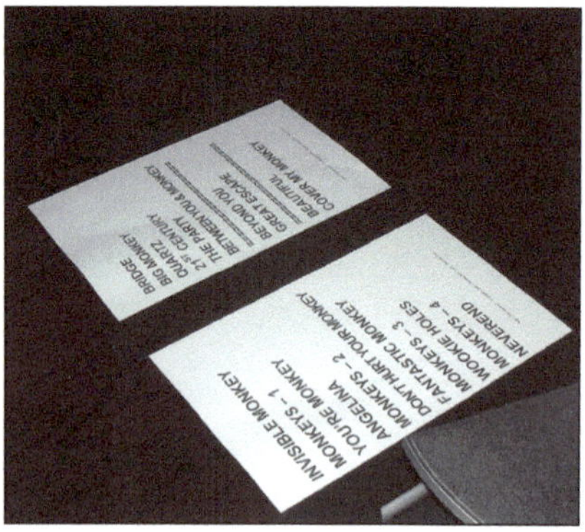

The Marbles gang (left)—Jodi, Denise, Rick, Cathy, Julie & me. And some monkey business with the Boston setlist (right).

The band were indeed firing on all cylinders. Everything just seemed to click on that tour. We met more fans that turned into true friends attending those shows—(Kim Robinson, Polly Peterson, Brian Steed & Cathy Loomis and Myron Grubaugh among others). We returned the favor to our UK convention chauffeur, Rick Collins, and hosted him for the New England shows. We went from city to city with ease and no travel mishaps. We attended afternoon acoustic shows and signing sessions in record stores. We had pre-show meals and after-show drinks. And it seemed in my memory like every day was filled with joy, wonder and delight.

Los Trios promo card (left) & laminate (right) and Boston setlist (center).

This high continued in 2005 when Marillion decided to squeeze in several more American shows before their *very* expensive U.S. travel visas expired. They returned to the states as a cost-effective three-piece ensemble, performing a semi-acoustic set billed as *Los Trios Marillos*.

Pete and Steve were accompanied by h singing and taking a full-time position at the keys, while a laptop filled in for Ian's drums ("*Laptop for President!*"). Here we were often presented with stripped-down versions of songs where h's voice was allowed to take center stage. Songs that previously showcased the power of the full band left us captivated with the beauty of a voice that is one of the finest instruments I've heard.

That's not to say we didn't get some rocking moments as well. At the Boston show, Steve Rothery put down his acoustic guitar for his electric and treated us to a full-blown outstanding version of "Sugar Mice" with that killer guitar solo that blew me away in the college dorms 30 years ago.

For the second tour in a row, Julie and I saw four shows—in Boston, Chicago, Philadelphia and (with no New Haven show this time) … San Francisco. (*Francisco…* that's fun to say.)

Story 16

Whew... Those were fun times. I'm thinking it's *break time* again. Sound good? Not much introduction needed for this one. If you're anywhere close to my age, then you were a Star Wars fan. If you weren't, then I'm guessing maybe you were Amish in the late 70's? Star Wars dominated our culture in a way few movies did.

If a kid can be so obsessed with a band that he instigates an industry-changing event for them, would it surprise you to find out that he would also be so obsessed with a movie that he would sit through three back-to-back screenings? Probably not.

Age 10, Post-Grade 4, 1977
Star Wars and the Ten-Year-Old Jedi Knight

Star Wars trading cards, Star Wars action figures, Star Wars posters, a Star Wars portfolio (kind of like a calendar of cool pictures), Star Wars bed sheets and even Star Wars PJs. I was a typical ten-year-old boy in 1977 when the greatest movie of my universe hit the theaters.

I saw the movie seven times on the big screen—once each weekend for four weekends in a row, then *three* times back-to-back on the same day. Alan and I were there with the Higgins kids, Steve and Scott. They were older than us and a bit on the wild side, so we loved to hang out with them. We saw the first matinee of the day on a Saturday long, long ago in a suburban town not too far away.

Star Wars-themed Marillion T-shirt promoting what would later become the epic double album, Marbles.

I was as enthralled with this (my fifth) Star Wars viewing as I was with the first. I'll never forget the opening scene when a rebel ship flies out from behind your viewpoint onto the screen. Then an imperial cruiser enters the scene in pursuit, laser cannons blasting. The nose of the cruiser soon takes up the entire screen but it just keeps coming and coming from behind the camera (the enormity of this vessel in contrast to the rebel ship a statement in itself). The gargantuan size of this thing (in a fantastic example of cinematic forced perspective) nearly snapped my fragile brain the first time I saw it with wide eyes and mouth agape.

But now, the matinee showing on this fifth viewing was over and we watched the people file out of the theater. When everyone else was gone the Higgins kids (those rebels!) said, "Let's stay and see it again!"

"What? See it again? Mom only gave us five dollars."

"Don't worry about it. Just stay here."

So, we stayed. And sure enough, the kid with the broom made his way over to us and said, "Show's over."

I knew we wouldn't get away with it, I thought.

But Steve Higgins (the eldest) was quick to respond. "We missed the beginning, so we're gonna stay, OK?"

The kid with the broom may not seem like much now, but he was definitely an authority figure back in 1977. I was shaking in my shoes. *He lied! Oh my God. We're going to jail. I know it.*

But the kid with the broom just shrugged and said, "Whatever."

Needless to say, the power of a lie was not lost on my little brain. It may even have been my idea to stay for a third viewing. I can't remember, but stay we did. I remember *that*.

Alan ran out to the parking lot again after the 2nd show and asked Mom (who had already driven *once* to the movie theater in an unsuccessful attempt to pick us up after the first showing), "Can we stay and see it again? Pleeeeeasssse!"

"*Three* times?"

Good old Mom. She let us stay.

This time I tried to take in the movie from a technical perspective. I remember staring up at the ceiling during the action scenes to watch the beams of light that traced the laser cannons and lightsabers. At times it looked like the theater was full of laser beams as they danced above our heads. But I couldn't keep my eyes off the screen for too long. Sure enough, I got sucked back in again...

Star Wars is *awesome*, I said back then. And I'm sure I'll say it again when I see the newly remastered version playing in a theater near you.

Thanks, George. You did good!

Story 17

As you've gotten to know me, you've probably picked up on the fact that I'm a fan of self-deprecating humor. I believe that if you can't laugh at yourself, you're wasting a lot of negativity on something you yourself are most able to appreciate and see the humor in. I love to smile as I look back and contemplate my former idiocy. (Well, hopefully *mostly* former.) This also has the added benefit of focusing my attention on any progress I might be making in becoming a better human. *Way to go, Jeff. You are soooo enlightened now.*

This following piece is in no way humorous (self-deprecating or otherwise). The thing I like about it though is this. I can remember so clearly being so convinced of my rightness and outraged at this injustice. Yet as I look back with the perspective of time and maturity, I now come to the exact opposite conclusion. (A perfect example of irony.) And that's what makes it interesting to me. Plus, you get introduced to Julie as a kid. And who wouldn't want that? Continuing our Star Wars theme, our next episode is called "The Julie Strikes Back!"

Julie and her nephew Tim Hanlon at Lake Tahoe.

Age 16, Grade 11, 1983
Julie Strikes Back

Well, it was lunchtime and we were all hanging out at our usual table in the cafeteria. Like most high school kids, we weren't popular but we had a circle of friends that we clung to with all the insecurity we could muster.

And sure enough, there was Chris, doing his usual shtick. He could be a jerk sometimes, but you got to admit it; he was pretty funny. And we all had to take our fair share of it. This time he was cupping his mouth to give it that muffled sound as he called out his pet name for Julie. I laughed along to the announcer's voice pronouncing "Rugby Woman!" and other nonsensical name-calling.

It was all par for the course. Nothing other than a standard lunch break on a typical school day. Until what happens?

Normally timid Julie erupts in a flash of frustration and the next thing I know, *I can't see*. For a split second, I don't understand what has happened. A moment later as I process the fact that I'm *drenched*, I realize an entire glass of water has just been thrown in my face.

My face! I'm soaked! *What? Why!?* Oh, the nerve... The *injustice*!

And what does Chris do? Ever the instigator, he adds on, "You're not going to let her get *away* with that, *are* you?"

Oh, no-siree. I looked left and I looked right, seeking anything to help me get restitution. (*I demand satisfaction!*) And there, right before me, are the remains of an ice cream rocket, melted chocolate cream still dripping off the plastic push-up base. I snatch it up and chase the fleeing offender. Over the table, around the corner until I've confronted her in the hall with not a teacher in sight.

I brandish my weapon in front of my assailant, but she's got her fear under control. With a slightly defiant raising of her chin, she prompts me, "Go ahead. *Do* it."

But there's something else there too, in the corner of her eye. I see what could very well be a tear that threatens to roll down her freckled cheek.

"Why the hell did you throw the water at *me*?" I ask with exasperation and lower the ice cream. "I didn't even *say* anything. It was *Chris*."

"You were laughing with the rest of them," was the reply. "I expected *more* from you!"

Story of my life.

I tossed the push-up rocket in the trash, unused. Good for you, Julie.

I am with you all the time now. One soul, one mind, one heart.

- Lyrics by Steve Hogarth

Chapter 31

Let's Get Pedantic

As you've probably surmised by now, I'm prone to digress and discuss the details of things that some of you might have little patience for. Did you enjoy that chapter about symmetry? OK then. Good. Let's get into the weeds once again and analyze some minutae. Was it a bit too esoteric for you? Not a problem. You might want to jump ahead a couple of pages in that case. We'll catch up with you in a few minutes. See you then.

OK. Now for the rest of us. Let's get pedantic…

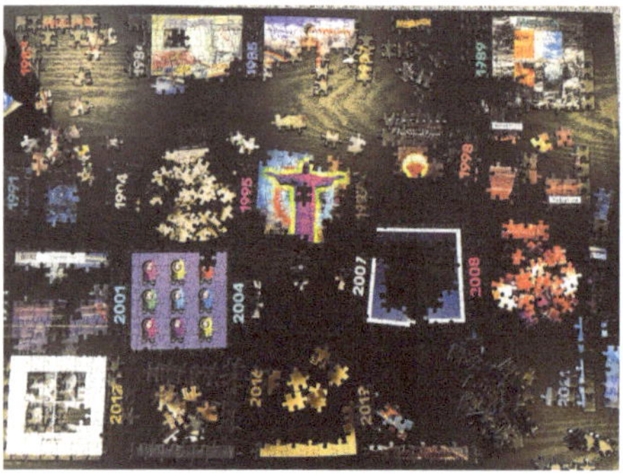

This was a pretty fun puzzle based on the Marillion "studio" albums. You can work on each album cover separately (*left*), and then work on combining them all in the correct order (*right*). But this puzzle has led me to a debate that has been discussed recently in the Marillion-verse… That being, how many studio albums have Marillion in fact released?

You would think this would be a pretty easy question to answer. Just count up the albums of new material, right? Well… not so fast. Let's start with the basics. I'm sure we can all agree that we can immediately exclude all Greatest Hits, compilations, Making Ofs and the many, many, *many* different live albums that have been released. So where does that leave us?

You can see from the 5 x 4 array in that puzzle on the previous page why the

Marillion camp is now stating that the upcoming album (*An Hour Before It's Dark*) will be their 20th. But is that really the case? How many studio albums have they *really* released? The new album will be the *nth* Marillion album. Let's solve for "*n*".

The absolute minimum value for *n* has to be 18.

- We have the four Fish albums—*Script*, *Fugazi*, *Misplaced Childhood* and *Clutching at Straws*.
- We have the four h-era EMI albums—*Seasons End*, *Holidays in Eden*, *Brave* and *Afraid of Sunlight*.
- We then have four indie albums that repeated the previously discussed symmetrical pattern—*This Strange Engine*, *Radiat10n*, *Dot Com* and *Anoraknophobia*.
- Five albums of new material followed those—*Marbles*, *Somewhere Else*, *Happiness Is the Road*, *Sounds That Can't Be Made* and *FEAR*.

All right. Let's add those up now. 4 + 4 + 4 + 5 = 17. Therefore ***n* = 18**. Right?

Again, that's the minimum. But is it correct? Let's discuss.

In 2004, Marillion released a double album called *Marbles*. A double album is *two* albums worth of new material on *two* separate discs. Does that count as two studio albums or just one? I vote for *just* one. If *The Wall* by Pink Floyd is considered one album, then so is *Marbles*.

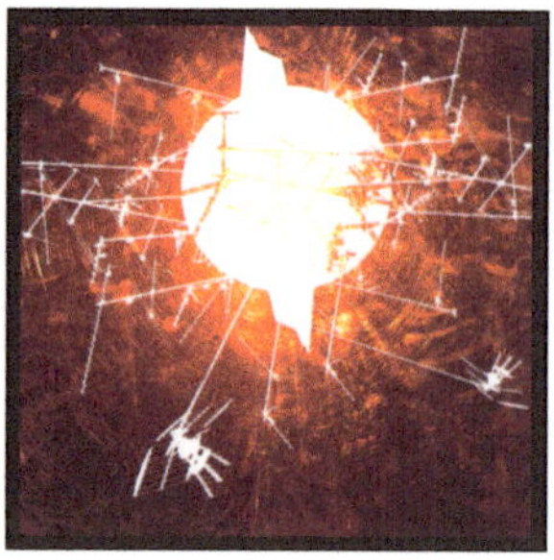

In 2008, Marillion released another double album called *Happiness Is the Road.* Same answer? Hold on. This time, not only were there two albums worth of material on two separate discs… They gave *separate titles* to each of those two discs (Vol. 1—*Essence* and Vol. 2—*The Hard Shoulder*). They were originally released in one package. Yet, at one point, you could buy those two albums separately on vinyl via the Marillion website. So now that we have two different album titles, is it one album or two? This gets tougher, but I'm still going to vote just *one* on *Happiness.* I bought it as a single package of music containing two discs. However, I accept the logic if you want to count that one as *two.*

In 2009, Marillion released *Less Is More.* In this album, they completely broke down a number of their previously released songs and re-recorded them with new instruments, often making songs that sounded completely different from the original version. It's the same lyrics and the same basic melodies to the songs. Yet haven't they created something new here? It's not just a remix. The music is new. Does it count as a new studio album? Again, I vote *no.*

In 2019, Marillion released the *With Friends From the Orchestra* album. Once again, we have an album of previously released songs. This time they have been re-recorded with a string quartet, a french horn player and a flautist. So, this feels a lot like the *Less Is More* debate. Same songs, but done in a new way. If I voted *no* on one, therefore I'll vote *no* on the other.

But not so fast. *Less Is More* isn't entirely re-recorded songs, is it? There's a new song on there ("It's Not Your Fault"). Does that mean we should reconsider our vote? It's a *new* song. I still vote *no.* Many singles are released with an unreleased track. In fact, they released an entire album of original studio material with the album *B'Sides Themselves.* And then did the same thing for the h era with the *Singles Box Vol. 2 '89 – '95.* These releases contained all of those extra studio tracks from the Fish-era singles and the h-era singles respectively.

And by the same logic of re-recording songs, we could argue that the *Remixomatosis* album would count. It's remixed versions of all the *Anoraknophobia* songs—many of which sound completely different from their originals. I think we have to stick with our original vote. Are these discs of original material not included on previous studio albums? *Yes*. Are they new studio albums themselves? I say *no*.

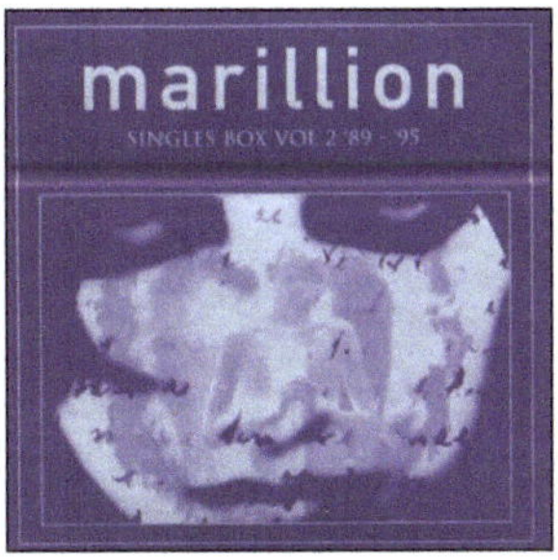

So here we are. Have we solved for *n*? In my mind we have. *An Hour Before It's Dark* will be Marillion's **18th** studio album. Do you disagree? I'm sure some of you do. Is the answer *n*=18? Is it 22? Or is it somewhere in between? The band says *n*=20. Send me your answer at marillionjeff@gmail.com. Cast your vote if you care!

Are we done with pedantics? Not quite yet. As I'm sure you can guess, I have my anal retentive tendencies. I keep Excel spreadsheets listing every book I've read, every CD I own and of course every concert I've been to.

What's my running concert total? Well, once again, that's debatable now. Before the pandemic hit, I could say with a pretty reasonable amount of certainty that I had attended 393 concerts and 35 theater events (including plays, musicals and comedy shows). I've seen shows in 13 different states and 6 different countries.

Do opening bands count as a separate concerts? Of course not. If the headlining band does an acoustic set at a record store signing session the day of the show, does that count? No. You could call that an opening set for that night's show. If you only bought one ticket, then that's still all part of one concert's activities.

Well, there is an exception. I saw Joe Jackson at the Somerville Theater on the *Heaven and Hell* tour in 1997. And we learned that Johnny D's (a club across the street) allowed people with theater tickets to pay no cover charge to enter. After Joe finished his excellent *H&H* set, we walked over to Johnny D's and caught a late show by Andy Summers of The Police. One ticket—but I count that as two separate shows—two different headlining artists at two different locations. Just because the second show was free doesn't mean it wasn't a whole 'nother show.

OK. So what's the issue? Well, here it is. Our favorite lead singer did an acoustic "*h natural*" show in the midst of the pandemic in 2020 at St. John's The Evangelist Church in Oxford, England. And he *live-streamed* the concert.

Did I attend the show, physically? Technically no. I was in my living room when we watched it. However, it wasn't like watching a concert DVD. I watched the show *live*, as it happened. I bought a ticket to the show and attained entry *virtually* to watch the concert. It was a private performance that we witnessed. Just because we didn't

travel to a separate location does that mean that I didn't "attend" a concert when I saw that *h natural* show, does it? Or more importantly, did I attend my 394th show or not? Does it go on the spreadsheet? Or does it not, simply because I wasn't rubbing shoulders with strangers in a location other than my home? These are the important issues that must be resolved.

Again, I will turn to you, my friends. Write to me at marillionjeff@gmail.com if you have a definitive opinion on this dilemma. Does the *h natural* live stream count as my 394th concert or not? Or do I simply include it with an asterisk? I'm curious to see what the collective voice has to say about it.

Thanks for your help. A few final answers to questions you probably never would have asked, before we rejoin the rest of our reading audience:

Hey Jeff, what about Marillion conventions? Three concerts or one? The answer is *three*. No question about it. Even though it's a single ticket, it's three separate performances on three separate days. The only way it would count as a single concert is if it was a three-day-long show that just went on and on like Woodstock. And *Swap the Band* counts as an opening set, not a fourth concert.

What's the all-time best non-Marillion concert you've ever seen? I have to cop out a little here and claim it's a tie. The two best concerts I've ever seen are the *Secret World Live* show by Peter Gabriel in 1993 and the *Radio KAOS* show by Roger Waters in 1987—both coincidentally at the Worcester Centrum in Massachusetts. My favorite Marillion show? Are you crazy? How could I ever pick amongst the fifty-plus Marillion concerts I've seen over the past 25+ years, including all those amazing convention shows? You're just being cruel now.

Are there any shows you regret seeing? Not really. There is only one show that I'm a little embarrassed to admit I saw, though. You now know my first-ever concert was the amazing Yes *9012Live* show (once again at the Centrum). My second concert though was Night Ranger at The Orpheum in 1984. I have to admit that I thought "Don't Tell Me You Love Me" was one of the best rock songs ever in my youth. I wasn't even there for "Sister Christian." I was there to *rock*! Ha. So yeah. The only concert I'm embarrassed to say I saw... Night Ranger. And now that I've had a little laugh at myself once again, let's get back to our main narrative.

Right after a philosophical question for the Marillion fans out there: If they play "Half the World," then follow that up with "The Other Half"... since they've played two "Half" songs, have they only played *one song*?

Chapter 32

Tour Fund Redux

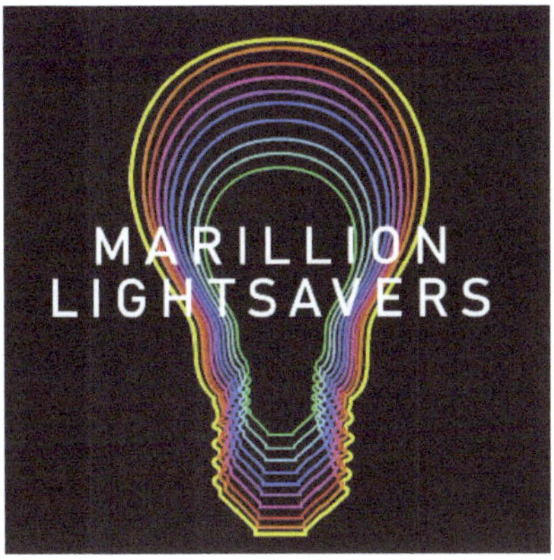

This book was supposed to be done. I wrote nearly 80,000 words and brought us all the way up to the present day. And then those damn Marillion guys did it again.

I've done a reasonably good job of staying pretty close to a chronological telling of all the events in our main crowdfunding story. Yet I want to save the last remaining chapters still to come for the end. So we're going to take a small leap forward to the present day before we return to those stories I wanted to close this book with.

2022 would have been the 25th anniversary of the original Marillion Tour Fund. And I had been doing some thinking about that. With the launch of this book, I wanted to open another Tour Fund to bring the boys back to the States for what very well could be one final hurrah on this side of the Atlantic. Every cent collected would have gone directly to the band to support their touring budget, regardless of how much or how little it raised. If we reached the "fundraising goal" however, Marillion would have promised to return to the USA and play a minimum number of shows in the next year or two.

To incentivize donations, the band could have done the same thing they did last time by giving a CD (or probably a download in these digital days) of one of the shows to everyone who donated a minimum amount. And to further incentivize donations,

I have two extra copies of the original *MARILLIONROCHESTER* Tour Fund CD from my extra donations way back when—the *ultimate* rarity. I was going to donate one copy to be given to a random person from the donation list. And I was going to donate the second copy to be randomly given to someone who pre-ordered this book.

Seems like a pretty good plan, right? Plus there was a certain symmetry to the whole thing that sounded kind of cool to me. The "inventor" of crowdfunding writes a book about crowdfunding and sells it through crowdfunding (by preorder). At the same time, he repeats the fund that started it all 25 years ago to bring back his favorite band to the USA one last time.

But as I said, those damn Marillion guys beat me to the punch. ("*Punch, punch punch.*" Ha.) And even though it's another Tour Fund, it's got a new twist on both why they did it and how they structured it. Let's start with the "Why."

We now know the new album will be called *An Hour Before It's Dark* and the November 2021 tour will be called *The Light at the End of the Tunnel* tour, consisting of 10 UK dates. If one of the band members came down with COVID, the remaining tour dates would have to be rescheduled or canceled. There is a huge financial cost in doing that. Normally, an insurance policy would be taken out in the event a band member gets hurt and the tour has to be scrapped. Yet in these halcyon days (he says sarcastically), insurance companies won't even touch that risk. It's just not possible. No one will underwrite it. So, if the entire tour gets scrapped, the band stands to lose ~150,000 Pounds. And there's no way to protect themselves against that possibility. That's the "*Why.*"

The "*How*" is as follows—introducing the "*Lightsavers*" Tour Fund. The band asked their fans if they were willing to step in as the underwriters for the unattainable insurance required to prevent the risk of the band losing their collective shirts on the upcoming tour. Everyone who donates will be receiving thank you gifts and raffle entries, the values of which are tied to whichever pledge level is donated. One coveted prize is actually free entry to Marillion concerts for life. Every pledge, however, will include a download of one of the shows and a downloadable "insurance certificate."

And here's the best part. If the tour goes on as planned, it all gets returned at the end. That's right. If the tour gets canceled, the fund will be used to cover those losses. But as long as catastrophe doesn't strike, each donor (including Julie and myself, of course) will be getting our money back in full. And the *Thank You* gifts will still be awarded regardless.

Pretty cool, wouldn't you say? Tour Fund Redux, but with a new twist—fans acting collectively as indemnification against calamity. Well played, Marillion. I wholeheartedly approve! (Not that you needed my approval, of course. Ha.)

Within four days of the original *Lightsaver* announcement, the fund had already collected over £110,000 of the £150,000 goal—completely selling out the higher £100 and £250 pledge levels. Eventually, the fund met its goal and the tour was on. (You weren't surprised to hear that, were you?)

Let's let the band themselves give you their spin on it, in their own words.

MARILLION - LIGHTSAVERS E-WEB

Calling the faithful......grab a coffee and have a read...it's a long one...

As you know, we are starting to prepare for our 'Light At The End of The Tunnel UK tour in November. To say we are excited about it is an understatement. We can't wait to get back out there and play to you all again (including the premiere of 'Be Hard On Yourself' from our album 'An Hour Before It's Dark'). We know the shows will be wonderful.

However......

As you know, the Covid pandemic is still out there and although we are all double vaccinated (as are our crew), it is still possible to catch it and (even though the vaccine protects from serious illness), according to current laws, if one of us tests positive whilst out on tour then we would have to isolate. Which, in turn, would mean the remaining concerts would have to be cancelled. This would be disastrous.

There is no insurance policy available to cover this. None, nothing, nowhere.

Although we will be taking every precaution available to us (even so far as travelling on a separate bus from the crew and not seeing our families or friends for the duration of the tour), we would still stand to lose in the region of £150,000 if the tour gets cancelled as we would still have to pay the crew, truck and bus companies, hire of lights and PA etc. This, along with the money that we wouldn't receive for the actual gigs not played, leaves us in a very precarious position.

You have been amazingly supportive with the pre-order for 'An Hour Before It's Dark', and that money has and will allow us to finish, manufacture and release the album early next year. Our fans are amazing. You helped us to invent crowdfunding and we need your help again.

In a world first, we would ask YOU to become our insurance company.

Everyone who pledges will receive thank you gifts and also the chance to win money-can't-buy prizes. At the end of the tour when we (hopefully) ha-

ven't had to cancel any gigs, all your money will be returned to you and you still get to keep the gifts. Obviously if we do have to cancel, then your money will be used with grateful and never ending thanks to pay the £150,000 bill we mentioned above. There are many different pledging levels and please only pledge if you can afford for the money to be used. We hope against all hope that we won't have to. And we also hope that you will enjoy being part of this, our most radical experiment yet!

But most of all—we hope to see as many of you as we can on the tour (though please don't touch!)

All details can be found below.

Thank you in advance to our 'Light' Savers.

h, Mark, Ian, Steve & Pete

A FEW WORDS FROM THE BAND...

Well, it certainly has been a tunnel.

We spent our time in it writing and recording our new album "An Hour Before It's Dark" which will be released early next year.

In the meantime we are very excited to emerge into the light with this UK tour.

I guess we've all - audience and artists alike - missed the vibration of live shows during this peculiar period in time, and we're pretty confident you'll be as emotional as we are to be together in a room full of colours and sound again.

We would like to thank you all for keeping the faith and being here tonight. You're the best, and we know it.

Enjoy the show...

HUGE thanks to anyone who got involved with our Lightsavers Insurance Fund.

Who else but OUR fans would donate so generously to insure a tour during a pandemic?!

Frankly, you made something happen which would be unimaginable for most bands. It's humbling to feel the love coming in, not just from the UK, but from all over the world.

Amazing.

ps We haven't gone to Rio with your money - it's safe and sound in a PayPal account and will be returned with thanks (fingers crossed) at the end of the tour.

Hugs all round

h, Mark, Ian, Steve & Pete

Extra special thanks to Shane McKie, who mailed us a copy of the tour program (which included this message from the band above, as well as the names of every single donor.)

The tour was an uninfected success, the gifts were delivered and all the money was returned as promised!

Chapter 33

A Most Unexpected Kindness

Those of you who know me would have expected a multi-page email recounting the following experience (if you had known it had happened that is—most of my friends don't). You see, at the time, it was simply too *special* to try to put into the appropriate words—words that would so surely fail to give it justice. Also, there's no way I could tell this tale without it coming off (to some degree) as "*Look How Great I Am*" bragging. And I cringe at the thought of this beautiful memory being tarnished by that unavoidable consequence.

Yet here we are, nearing the end of our time together. And what kind of storyteller would I be if I didn't tell you about the conclusion of this narrative? ...the moment when Mr. Holland hears his opus, the moment when Nash is given his mathematician's pen in *A Beautiful Mind* (such a satisfying movie-going moment that it literally made me catch my breath with emotion when I saw it on the screen for the first time)?

A few years have now passed since this all went down. And I am only now ready to attempt to distill this bit of my charmed life down into language that can be passed on to you, my friend. (Forgive me if I just assume that we're friends by now. That's kind of what friendship is about; understanding each other, right? And I have a feeling that you have a pretty good understanding of what makes me tick at this point.) And so, I delve into my memory palace once again to bring you a moment in my life that will forever be remembered as perfection in so many ways.

You know by now that things tend to work out for me. Many of my friends often revel in or resent the many fortunate things that tend to happen to me. My good friend, Dan Peduzzi (he of the life-saving ride to my doctor's office), honestly believes that there is only so much luck in the world. Therefore, he needs to chop off my head *Highlander*-style to absorb my unfair share of it. Well, here we go again with another example of too much good heading my way. So do me a favor; watch my back for tall skinny nerds carrying swords...

In 2018, Marillion toured the US again. I say *again*, but this is not a common occurrence, as we've established. This time, however, they made some odd choices about what cities they would play. They decided to forgo many of their tried-and-true locations and play cities that they hadn't played before. So not only were Boston and New Haven not on the table, we couldn't even travel to NYC or Philly to see them. The closest shows were in Buffalo or Greensburg, Pennsylvania, of all places. Hey, nothing against those towns. I'm sure they're both very fine locations, with many enjoyable things to do and places to see. It's just an unexpected choice for the *only two* shows in the entire Northeast corridor of the U.S.

We'd never missed a Marillion tour since that wonderful first night at The Paradise way back in 1995. In fact, when they come to the States, we usually see several shows. Recall that on the 2004 *Marbles* tour we saw four shows, driving to Boston and New Haven and flying to Chicago and Philly. I looked at the current tour schedule and thought the last show in Dallas might be the best option since…

Let's check the boxes. It was a city we'd never been to. *Check.* It was on a weekend. *Double Check.* And it would have relatively warm weather during the cold of a New England winter. *Quintuple Check!* We couldn't commit right away though, as Julie's work schedule was pretty tight at the time. Good thing we didn't book it. As it turned out (and as we expected), Julie wouldn't be able to make the Dallas show.

But with only a couple of days' notice, Julie called me from her work on a Friday and said she was sad we were going to miss the Marillion tour. Then and there we decided the Monday show in Durham, North Carolina, could work. I said to give me a minute and see if I could pull it together.

I found a name-brand hotel right next to the venue with reasonable rates. On the Ticketmaster site, a few primo seats had just been released. I was able to get us face value (*not resale*) tickets in the orchestra pit—one in the front row center (*!!!*) and another in the second row a couple of seats over. OK. Now we have to see this show.

And best of all, we had previously talked to Julie's niece, Amanda Hanlon, who worked as a flight attendant. She had already offered to give us a couple of her "Buddy

Passes" when we talked with her about going to Dallas. These are free flight vouchers that airline employees can give to friends and family. Yes. You heard me. Free flights. (*Sweet!)* Like I said, things just always seem to work out for me.

Amanda scoped out the best options for our updated destination. It was going to be tight. A lot of the flights were close to being sold out. (And you're the first to get bumped off a full flight with a Buddy Pass). But she made it work and booked us our no-cost travel. Our entire trip came together perfectly with only two days' notice.

As we get further into this story, never forget that Amanda is the hero of this chapter. Because if last-minute flights had cost too much, I may have ended up being too cheap to pay the high prices. I know myself. And while I love to occasionally splurge on myself, I also *detest* wasting money.

We flew down the day before the show, filling the very last open seats on the plane as expected. Once we arrived, we discovered two of our favorite people in the world, Kim Robinson and Polly Peterson, were also in town for the show and were staying at the same hotel. (Shout out to two of my favorite Marillion gals!) That night we had such a delightful time going out for dinner and drinks with our friends at a fantastic tapas place, that the trip would have been completely worth it even *without* the concert we were all there to see. (*Kim, Polly, you guys are truly wonderful people and I'm so glad we've gotten to know each other over these fun-filled years.*)

The next day had some fun moments. We ran into keyboardist Mark Kelly who stopped to chat with us for a minute. We told him about our last-minute decision to come see the show and how easily it all came together. I went out for ice cream shakes with Polly while Kim got a Reiki session from Julie. I went on a Durham Bulls memorabilia quest with Kim while Polly had her turn with Julie's magical powers.

Afterward, we also saw vocalist Steve Hogarth in the lobby and Julie ended up giving him a quick Chakra balancing right then and there—a session in which he started to really open up about himself. So I said, "*Let's give them some space*," to the people in the lobby who were watching in fascination.

Eventually, it was time for the show. We settled into our seats up front, with Julie right below h's mic stand and me just behind her and a few seats to the left. How great to have excellent seats reserved for us at a Marillion show instead of the usual *General Admission* and standing on aching feet for hours, where we short guys are stuck just seeing the tops of the band's heads?

They opened with "El Dorado"—a 16-minute epic from the latest album *FEAR*.

For me, the dense and challenging tunes can often take time to fully seep into my consciousness. Sometimes it's not until the song unfolds in a live presentation that I can finally *really* appreciate a song like this. And that was the case with much of *FEAR*. The outstanding synergy of the video presentation with the lights and sound just blew me away. My appreciation for this tune really started to gel and we were off to an excellent start to what I knew was going to be an exceptional night.

All of the self-awareness had seeped out of me as (in the darkness), I sank into the anonymity of the crowd. I just sat back and enjoyed the moment and my good fortune. How wonderful is it to be able to put together an experience like this—seeing a new place, connecting with old friends unexpectedly, seeing incredible music performed for you from such a great vantage point and having it all come together with such ease? Life has indeed been good to me and I truly appreciate it...

As the customary thunderous applause washed over the band, Steve H said his usual hellos and thank you's to the crowd. But before he introduced the next song, he suddenly said, "Just before we went on stage, I heard Jeff Pelletier was here and I'd like to dedicate this *entire evening* to him... wherever he is."

At this point, I felt somewhat disconnected from my corporeal self. It was like a couple of gears inside my brain had skipped a few cogs. I immediately thought, *I must have just imagined that, right?*

My jaw must have dropped open as I sat there stunned. He covered his eyes from the spotlights, and scanned left and right through the crowd. At this point, Julie started to point me out behind her to the left. And another long-time friend of ours, Jodi Leigh (*Hi, Jodi!*) was further down my same row to the right. She also started to animatedly point to her left.

I half stood and waved, not sure what would come next. H locked eyes with me and then with a big grin on his face, ran over to the side of the stage in front of me. I thought he meant to shake my hand, so I stepped up to the stage and was unexpectedly embraced in a bear hug.

At first, I thought maybe the crowd was just applauding because that's what you do when you're at a concert. *He said to applaud, so we're applauding.* But as h continued to hold me tight, I heard the crowd behind me get louder and louder as the embrace continued for just a few extra seconds longer to convey the genuine, heartfelt affection in the act. Eventually, he let go and I turned around to see people not just applauding, but on their feet *cheering*. I even heard a few people yelling out my name.

I honestly can't properly describe the emotional response this evoked inside me. I felt as if my chest was physically expanding as this incredible gratefulness welled up inside of me. This was so unexpected. And the response from the crowd, so utterly enthusiastic.

Humankind can be just that sometimes, so kind. And I had one of those moments of unblemished, unequivocal happiness. I'm not ashamed to admit that my view blurred ever so slightly as a sheen of water filmed over my eyes.

Not knowing what to do, I gave a little bow to the crowd and the noise level notched up once again. I looked out over the crowd one last time. From left to right, from down on the floor to up in the balcony, a thousand people on their feet, sending this message of love and appreciation. *Thank you.* Thank you to every single person in that audience. Thank you for a perfect moment in a life already so full of fortune and joy.

All of this of course only lasted for seconds, even though it seemed like minutes in my memory. Meanwhile, h ran back to the mic stand and said, "That's the man who invented crowdfunding and we took all the credit for it!" Another surge in the roar from the crowd followed.

Then as if that wasn't enough, he added, "There's a genius in our midst, ladies and gentlemen."

I have to admit that I was completely overcome at this point. I had already re-taken my seat and bashfully put my hands over my face, laughing into my palms slightly with this gratified emotion coursing through me like a river of positivity within my veins. Thank you again.

Thank you, whoever you are
Thank you, whoever you are
Thank you.

- Lyrics by Steve Hogarth

Chapter 34

As if That Wasn't Enough…

The concert resumed. And I have to say, it was a great one. Before the show started, I had a nice conversation with the friendly guy sitting next to me. Well later, after "The Invisible Man" brought the house down, the Friendly Guy leaned over and yelled in my ear, "Dedicated to *You, man*!" Ha. *Nice one, Friendly Guy*. I'm sorry I forgot your name, but you're a cool dude.

Eventually, the lights came up and Julie made a bee-line over to me. One of the most amazing things had happened to me and because we were seated apart, we hadn't had a chance yet to talk about it. I gave her a massive hug and simply asked, "Can you *believe* that happened?"

But as we turned around, to my amazement, we saw people starting to line up to talk to me. Some familiar faces, and some brand new to me. As if this crowd hadn't already given more than I could ever express my appreciation for to them… I now shook hands and talked to a number of fans who wanted to meet me, wanted to thank me, wanted to grab a picture with me. Do they know I'm not a celebrity? Apparently not.

I talked to a number of fans down in the orchestra section. Then we worked our way back to the lobby and the same thing happened again. I wound up talking to a circle of fans, again not understanding why everyone seemed so eager to meet a regular guy from New England who lives a life of quiet anonymity.

This would have continued even longer than it did, but Julie eventually came over and grabbed me, saying they were about to shut the doors to the backstage entrance. Marillion management, Lucy and Stephanie, had been kind enough to include some backstage passes with our tickets after I sent them an e-mail telling them of our last-minute plans to see them. So, we made our way to the after-show.

In a small venue like the Durham theater in NC, an "after-show" consists of a single room that holds a few dozen people. On one side of the room are two tables loaded with snacks and drinks. The band members were here, each individually talking to friends and fans and relaxing after a night of emotional catharsis in the form of their music.

Over the years, Mark Kelly has always been steadfastly nice to Julie and me. He even mentioned me by name in his TED talk. We made our way over to him and said hello. I had to ask him, "That was the coolest thing that has ever happened to me. Do I have *you* to thank for that?"

Mark replied. "Oh, no. That was all Steve." He then told this charming story of how he and h had talked before the show.

Do you know who's here tonight?
Sure. Jeff Pelletier's here.
How did you *know he was here?*
I saw them earlier today. How did you know he was here?
I saw them in the lobby.
He and Julie are sitting in the first and second row...

Another delightful memory. Being told that two of my idols from my favorite band were almost bragging to each other about which one knew first that I was in attendance that night. Now let's go back to that fanboy I used to be back in college. Imagine if I could give him just an inkling of what would happen to him this night 30 years later. And how much would he love hearing *on top of all that*, that the band were actually talking to each other about me being there to see them? I never would have believed it.

Chapter 35

We Haven't Even Gotten to the Best Part Yet

Naturally, at this point I had to thank h for what he had done. And there he was with that impish grin on his face, chatting to those hanging on to his every word (as those of us all do when in the presence of someone we admire, idolize or just plain recognize as having that singular undefinable thing we call artistic talent).

He was wearing a blue shirt and eating blueberries (as Julie had suggested he do during her Reiki session with him earlier that day). "I do as I'm told," he said with a knowing grin when she pointed out that he was focusing on the recommended *Blue* color.

Julie also thanked him for playing "Beyond You" that night. She had made an offhand remark during their chakra balancing earlier, that she'd be fine as long as they didn't play "Beyond You" tonight. It was her favorite song, and she knew she'd break down in tears if they played it. H's kind response was, "I played that for you."

Here's where this story gets hard for me to tell. I've already mentioned that I am not comfortable even alluding to self-congratulatory statements. So, I will now switch to the omniscient third-person narrative voice so as not to cringe at the praise I'm about to re-tell. It's still going to be hard for me to describe, but I want to share this with those of you that want to hear it. So take my hand and let's go together…

Jeff thanked h and related what it meant to him to be recognized in such an unexpected way. H then proceeded to tell Jeff that he felt that it was long overdue. He told him that all of the band feel that what Jeff did for them was a career-changing act for them. If he hadn't shown them how they could use the internet to succeed with such a passionate (but relatively *small*) fanbase, then they never would have achieved all that they did and there's a good chance that they might not even still *exist* as a band today. He further said that they all feel that in a way they owe their *entire existence* to him.

Needless to say, Jeff was stunned to hear such a strong declaration of gratitude delivered to him in person and stated so matter-of-factly. He then responded about how if there ever was a band that deserved to be saved, it was Marillion. He doesn't remember everything he said in that conversation. Yet he does remember a small group of fans forming a little circle around the two of them as they talked about the importance of music. Jeff unspecifically recalls describing how there is something so inherently *satisfying* about listening to sad music. And he remembers h nodding his head in agreement, saying *because it's true.*

They talked for only a short time, but Jeff stepped away from that interaction once again feeling like he was living someone else's life. One incredibly nice fan then introduced herself to him and actually stated that she felt honored to have witnessed that conversation. She asked for a picture with him, which of course, Jeff agreed to and spoke briefly to her and a few other fans. One of those fans was a musician who said he was so happy to get a chance to meet Jeff and thank him. He told Jeff how his band had crowdfunded their first album based on what they saw that Marillion had done. So, he said, if it wasn't for Jeff, his band might not have ever made that album—another example of how far-reaching our deeds in this world can be.

Jeff didn't speak to Steve Rothery or Ian that night (though both of them had at previous times exchanged kind words with Jeff and Julie). Jeff and Julie did talk with Pete Trewavas. Like Mark, Pete had through the years always made a point to say hello to Jeff and Julie whenever he spotted them after a show. During his Transatlantic tour he once even broke away from a crowd of fans in NYC to say he wanted to *say hello to a couple of friends* after seeing the two of them over to the side.

Pete was his usual gregarious self with a big smile and a welcoming demeanor. Jeff thought that he had already heard his quota of generous superlatives that would be spoken about him that night, but Pete added one more gem to Jeff's overflowing satchel of praise received.

Pete noted that what Jeff did wasn't just a *generous* thing. He said it was an incredibly *intelligent* thing. And people may think it's nothing special now, but it had never been done before. And genuinely smart people come up with ideas that seem obvious after they've thought them up. He then said he'd always be grateful for that.

Once again, Jeff was nearly speechless with appreciation. It was like someone decided to make a movie about a man having the time of his life—a man who never saw it coming and was floating through an evening of fairytale-like magic.

With h's kind permission, I can include the following words that he shared with me later regarding the dedication that night.

A Note From Steve Hogarth

As for the shout-out, it was long overdue. I don't think it's an exaggeration to say that without your kindness, cleverness and inspiration, Marillion's path may well have been one of decline and disintegration. (Who knows?...) You certainly planted the seed that made us free to create the music we have these past years. For this I consider us to be permanently in debt to you, (and how lucky to be permanently in debt to such a decent chap).

Thanks again for everything, my friend.
h
x

Chapter 36

The Icing on the Cake

Chatting with h at the next Marillion convention in Montreal, out of the blue he said, "I want to send you a birthday card." After telling him my birth date I pointed out that it was easy to remember if you know it's the same as Bob Dylan's. To which, he proceeded to spontaneously serenade me with two verses of "Subterranean Homesick Blues." "Johnny's in the basement, mixing up the medicine..."

The above card (left) arrived shortly after the next time my age ticked up on the odometer once more. Thanks, h! Was this kind offer the initial spark of inspiration that evolved into him sending personalized birthday cards to all "Purple" patrons of The Corona Diaries *podcast? (See Julie's card on the right.) I can't say. But it's a wonderful tradition he started.*

The night ultimately had to come to an end at some point. And so, we eventually made our way to leave the backstage gathering. We said goodbye to several of the people we had met. H was by the exit and so we stopped to say farewell.

To my surprise, h then asked if I wanted to see another Marillion show. I told him about how we originally wanted to see the Dallas show on the final night of the tour, but Julie had to work. I then wondered out loud about whether I should make a last-minute solo trip since this night's show had been so damn good.

H then said I should write to him "at this e-mail address." He said if I ever wanted tickets to a show, or merchandise or *anything*, all I had to do was write to him. He said he would consider himself forever in my debt. So, if I wanted to see a show or have him write a message to someone I love, or *anything at all*, he would consider himself on call to me anytime.

Julie and I have such a great relationship that we jokingly give each other shit all the time. And in that moment, she couldn't help but roll her eyes and say to h kiddingly, "Hey, I've got to go home with him!"

I will always love that she made that joke in that moment. Because, in response, h then put his hand on her shoulder and with the most earnest sincerity he could summon, he said to her, "Take care of him. *This is a great man.*"

A Note From Julie

For those who know Jeff and me, spontaneity isn't really our thing. Between my hatred of change and Jeff's indecisiveness, we tend not to make decisions quickly or on the fly. So that is why the trip to North Carolina was in itself a miracle. We made the decision to go on a Friday, (three days before the concert). That is insane. I didn't have enough time to truly stress, worry and get anxious. I might just go and have a good time. What? Unheard of. From the second we got to Manchester Airport, I think we both knew something special was happening but we had no idea what was waiting for us.

I will skip ahead to the concert, and the moment when h brought up Jeff's name. When h ran over to give Jeff a hug, the sound was deafening. There was an actual roar from the crowd. I turned around to look at the stage and noticed each of the band members were also clapping with big smiles on their faces.

After h started singing again, I turned around and looked at that face I love so much. He was in his own world. He was so happy. And yeah, I saw tears. It was truly an honor to be witness to someone feeling all that love.

Now what Jeff didn't portray in his retelling of the story is how I became his personal assistant for the evening. As he was being mobbed by the fans, I was constantly being asked to hold things, and fans handing me their phones asking me to take a picture of them with Jeff. I actually had a fan say to me, "Did you know that is Jeff Pelletier? Have you ever spoken to him? How cool is this?" Part of me wanted to sarcastically say, "Yeah. I'm going to ask him to sign my boobs."

As we finally got to the after-show party, I will share another little tidbit that I find so endearing. In this room, they had food out, and every kind of alcohol you can imagine. Band members and fans were drinking wine, beer, vodka, scotch. But if you know Jeff, you know he isn't much of a drinker so he asked if I wanted to "split" a beer. Well sure. We are on vacation. Why not? Who knows, maybe we can even "split" another one. I didn't have to share the first beer though, as Jeff got truly excited because he found juices in the fridge. "Look, Julie, they have JUICES!" Drink away, my friend. Drink away.

I was standing next to Jeff while he was talking to h. It was such a genuine conversation. H was truly so interested in talking with Jeff. And Jeff shared something with him, that I will never forget. While I am no Dr. Phil, I do know my Jeff. I have known for years that processing emotions was always a challenge for him. So he turned to music. I think that is why he always had such a strong love for music lyrics and live shows. It allowed him to express emotions in a safe manner. But during this conversation, Jeff thanked h, for the music, because Marillion's music allowed him to grow up emotionally. The emotions that h will sing with and about, allowed Jeff to feel and express. And that sentiment really appeared to touch h. So Marillion helped Jeff grow up. Helped him come in touch with his emotions and allowed him to let others in.

So yes, this was a night to remember. I am always proud to be by Jeff's side. But this night I had never been so proud of him. He handled it all so beautifully and allowed himself to feel it all.

Now that being said... when I did say to h, "Come on, I have to live with this guy," I did mean it like... Enough... His head is big enough, I have to go home with him. I meant it as a funny jab, but I am the one who got the jab as h said, "Take care of him, this is a great man."

Now ain't that the truth.

Chapter 37

One Last Joke

Later that night as we lay in the darkness of our hotel room, we talked about the wonderful night we had just experienced. Back and forth we relived so many moments from a night that would have been wonderful even if it had only contained just *one* of those memories. And yet they had all happened in the span of a few brief hours.

After a few minutes of silence meant we were both heading towards happy slumber, I asked one more time, "Remember when they played 'Beyond You' just for you?"

A moment's contemplation in the darkness was followed by a happy "*Yes*" in response—spoken in that way you know contains a big smile, even if you can't see it.

Then after another moment, I asked, "Remember when they played *every single song* just for me?"

I really enjoy being a shit sometimes.

Chapter 38

Choice Awareness Protocol (aka "Taking Back the 90%")

OK. So that night in NC was the event that prompted me to write this book. I hope I did it justice, because it really did mean the world to me. And now we've covered just about everything I had to say, everything that essentially led up to that night. So, is it finally over, Jeff? Are you *done*… finally? The answer is "*Almost*…" If you're interested, keep reading past the "sign off" to learn a few of my favorite recipes (and a way to lose some pounds if you enjoy some of those dishes a little too much). Before we get there though, there's one last topic I want to talk about with you. Humor me for just a few more pages…

Speaking of recipes, what if there was a recipe for a better life? I know that sounds contrived, but if you had the opportunity to be a moderately happier person on the whole, wouldn't you want to take advantage of that? If you would (*and again, who wouldn't?*), I want to share something with you. Don't worry. I'm not going to throw some New Age mumbo jumbo at you. I want to talk about action and result. It may sound a bit crazy at first, but I had some real success with it. So, if you think you could use a little infusion of positivity into your life, give it a try. (Or more accurately, an extraction of a little *negativity*.)

Think of a time when you were in a bad mood and you had to take that attitude into a public forum. Maybe you got in a fight on the drive over to a dinner party at a friend's house. Then you have to walk in together and act like nothing is wrong. Have you ever noticed that when you pretend to be in a good mood (even when you're definitely *not*), after a short of amount of time you realize that you're not pretending anymore? That's because attitude is programmable.

Now I know there isn't a switch inside our brains that we can flip and instantly change anger to joy, or stress to calmness. Nor do we have lines of code in our brains we can access and edit. We can't change the *If-Then* commands that determine our responses to stimuli. But what if you actually could program yourself (through long-term conditioning) to react to things in a more positive way than you used to? So you didn't have to overcome unnecessary negativity to begin with. Wouldn't everyone want that ability?

A bit of background first. I was a pretty tightly-wound kid. When I think back on my childhood, if you ask me to sum it up in one word, I know the perfect one. That word is *angst*. Why is that? Well, it's not easy to talk about. Additionally, I've tried to avoid having my words about the past come across as a cry for pity. Because of that, I don't want to get too far into the specifics of it. So let's just sum it up by saying that I grew up around a lot of anger. What seemed like a constant rain of antagonism and verbal abuse had pelted down on me, day after day after day after day. Never feeling safe from the potential threat. Never experiencing a moment of happiness not tinged by worry and angst.

When you grow up around anger, it pervades your psyche. And sure, the anger wasn't *always* on display. Yet the *worry* of the unleashing of that anger was always in the forefront of my mind. So as a child I constantly walked on metaphorical egg shells —always afraid to poke the tiger and never being able to fully relax, or just be unconditionally happy in any given moment.

Living with that kind of ceaseless anxiety shapes a person. And unfortunately, I was powerless to avoid that outcome. Those of you that are Marillion fans, would not be surprised to learn that the song "An Accidental Man" resonated with me instantly, with that line "*I was taught from much too young, to never give myself away.*" My introversion had been figuratively beaten into me. But beyond that, I realized that I, like many other kids, was *taught* my negative behavior and self-image by my life experience in those early years.

There's a pretty broad swath of interpretations that could be applied to that lyric. Yet I heard it and knew exactly what it meant to me—even if it means something completely different to someone else, or even the man who wrote it. To me, it immediately rang true of how hard it was for me to learn how to open up and share, to be able to connect, after a life of being shouted into my figurative withdrawal. Happiness was indeed "not a thing I ever learned to do" properly. *(Note to reader: You don't have to feel bad for me. It gets better. I promise. Just keep reading.)*

These chains are all your own, these chains are comfortable
This cage was never locked, born free but scared to be
This cage was meant for you, with care and constant attention
This cage is safe and warm, will you die and never know what it's like, outside.

- Lyrics by Steve Hogarth

Around the time I turned 40, I made a conscious decision that I no longer wanted my negative reactions to the world around me to be second nature to me. I didn't know how I could do it. But I wanted to make an effort, to at least try to break out of the pattern of behavior. I just finally wanted the past to be the past, and I wanted to minimize the unnecessary stress I so often felt. In short, I wanted (generally) to be a happier person.

Life can throw some big obstacles in your way. We all encounter immense losses over the course of our years. Some of those losses at the time can seem almost insurmountable. Let's face it. Some shit is just too big not to feel its effects. That's the *Serious*

Shit stuff—death, loss of job, loss of relationships... Yet most of life isn't the big stuff. It's the small choices we make every day over (mostly) inconsequential things.

I'm sure it's been said before in many ways. I look at it like this. Life is 10% what happens to you and 90% how you react to it. You can't really do anything about the 10% except try to get through it and survive the serious shit. But that big honking 90% leaves a lot of room for improvement.

So again, most of the time we're not dealing with those big challenges. Often what bothers us are the little things—the 90%. And if we could have a better attitude about the 90% (even just an incremental improvement), wouldn't that have a great effect on our overall well-being?

Now we're not robots. I understand that. But if we can re-program our mood at that party, why can't we re-program our attitude towards other things? I said I wanted to try something. So, I decided to conduct an experiment—on myself.

Whenever I found myself getting pissed off or stressed out, this is what I decided I would do. I would allow myself to feel that negative emotion. Again, there isn't a convenient OFF switch. So, I would still *feel* it, but from that point on, I would also say to myself, "This is a choice." Every time I got annoyed about something inconsequential... *This is a choice.* Every time I got stressed about some artificial deadline that I had forced upon myself... *This is a choice.*

I did this for years... literally. Over and over again I kept reinforcing the concept in my brain that I was *choosing* to react negatively to these minor things. *This is a choice.* Eventually, it became something I did without even trying. It became second nature to me to say, "*This is a choice.*" I continued doing it year after year. At first, I couldn't tell if I was accomplishing anything. I kept doing it but didn't really think about it anymore.

And do you know what happened? After about a decade I looked back and realized that slowly but surely it had worked. I found that (without realizing when it had happened), at some point, I had succeeded. Little things that would upset me before, I was now laughing off.

The experiment was a success. Over the course of years, I had actually conditioned myself (or as was the original goal of the exercise—reprogrammed my personality) to a small degree. At heart, I believe I was always a decent guy. But could people see that? I'm guessing not as well as I would have liked. Why do I say that now? Because I notice now that people are not just more receptive to my friendship, they actively comment on my likeability. And that did not happen before. Therefore, I felt internally that I had shifted my natural resting state toward the positive. Additionally, I felt externally that people had noticed the improvement.

Maybe that's an unrelated effect of just growing older and mellowing out. It could be that I'm overstating the degree of my success and confusing correlation with causation. But no. I honestly think I achieved my goal. I feel generally happier now. And I believe that people have noticed the change.

I do believe that this methodology works. Maybe it already existed and I had just never heard about it. Or maybe I've invented a self-help technique that wasn't around before. It doesn't matter, because I'm not going to try to lay claim to it or profit from it in any way. I just want to tell you all about it in case there's anyone reading this who might benefit from it.

How do we summarize this? Well, why not try to apply the scientific method to re-programming personality? Again, unconventional, but let's do it.

- Purpose/Question: Can we program our emotional reactions and permanently alter our behavior patterns for the better?
- Research: We can do it in the short term (like pretending to be in a good mood at that party). Let's see if we can do it in the long term.
- Hypothesis: By reinforcing the *awareness* of my choice to be annoyed or upset about something, can I (over time) condition myself to make a different, more positive choice?
- Experiment: Set up a protocol in which I state "*This is a Choice*" every time I feel a negative emotion. I can't clone myself to create a control group. All I can do is hope that I'll eventually be able to recognize the effect if the protocol works.
- Analysis: After many years, I finally realized that I was indeed reacting to things in a more positive way. The change was too slow to notice at first, and too incremental to be able to differentiate its progress in stages. Yet with the long view of many years, I'm convinced the experiment was a success.
- Conclusion: We can in fact succeed in re-programming our unconscious reactions to be a generally happier person.

Some consider Communicate to be an additional, final step in the scientific method. I think you communicate your results after all is said and done. So, I would argue against it being the final step. But we're communicating here now, therefore we've done it anyway. So why quibble about it?

Remember what we said about Life and the 10 and 90%? Let's put that into equation form.

$$\mathbf{L} = (0.1 \times \mathbf{H}) + (0.9 \times \mathbf{R})$$

Where **L** is Life, **H** is what happens to you and **R** is how you react to it.

Nerdy I know, but math is fun. And algebra is complex math, so it's compound fun (even if that's only a linear equation). Think of the weighting of the two variables in that equation. If it's anywhere near accurate, then a small change in **R** has a great effect on **L** (much greater than how you deal with whatever **H** comes your way). So even if you only shift **R** towards the positive incrementally, you will achieve a significant benefit in your **L**. I know I have.

As I write this now, I can say confidently that I'm a happier person than I was before I did this. I'm not a *Stepford Jeff*. I still feel anger, stress and all the negative emotions that are tied to the human condition. (If there wasn't any bad, we could never appreciate the good.) Yet my resting state has my emotional needle tipping more towards the positive than it used to. I naturally smile more. I appreciate how good it is to be alive and healthy and happy. I just feel *better*. And people respond to that.

What should we call this methodology? How about the *Choice Awareness Protocol*? (I tip my CAP to you, sir.) Let's do it. I just made that the title of this chapter. I've written this procedure here in the hopes that if someone out there follows the established protocol, they also have similar results and improve their overall mental state of being.

And here's another Marillion tie-in. Shortly after I started to do this, Marillion released a song called "A State of Mind." In it, h sings about how "*a state of mind is a contagious thing*." How true is that? Emotion truly is like a contagion. You can infect people with your anger and it will pass from you to the next person you snap at. Then they take it out on the next person they interact with, perpetuating the chain reaction of negativity.

But happiness is also infectious. Think about how a genuine smile from a stranger can lift your mood. And how you can pass that mood on to the next person you meet if you choose to... Well, my goal is to infect people with positivity. I don't care how corny that sounds. I'm doing it and you can't stop me.

A state of mind is a contagious thing
Spread it around, you never know what the future brings.

- Lyrics by Steve Hogarth

I used to carry a lot of guilt around with me. As I was growing up, I was told over and over that with my intelligence, I would be able to change the world. I would discover the cure for a terminal disease or invent something that would revolutionize our lives. As I worked for years in the field of electro-optics, I kept being hounded by that little voice in the back of my mind. *You're not changing the world, Jeff. You're selling out.*

Over the years I had a knack for seeing and understanding how things worked. That combined with an excellent memory always enabled me to test very well. (This probably made my teachers think I was smarter than I actually was.) I remember taking tests in which I would actually visualize myself taking notes in class and watching myself *in my memory* write what I needed to remember in order to answer a test question correctly.

Those days are in my past now. My memory isn't what it used to be. Did I squander what talent I did have for nothing more than financial gain at the time? In a way I did. But as I grew to become a happier version of myself, I came to recognize something. I may not have changed the world in the way I might have hoped that I would have. I haven't changed things on a global scale, but I have changed them on a local scale.

As I stated before, I am striving to be a source of positivity to color the world around me with joy. I am spreading happiness like the contagion that it is. I have made it a point to tell those that are closest to me how much they mean to me. I am trying to be kind and generous with those around me. I want to be a beacon of positivity that lights my tiny part of the world and have that light reflected beyond my reach.

And nothing is a better living embodiment of the success of this mission than the person I spend my life with. Over the decades, the love and assurance I have poured into this relationship have helped her to grow into the confident, joyous person that she is.

I was once an arrogant, insecure kid who thought he knew everything. Julie was once so self-doubting that she thought she didn't deserve to have friends. Now, though, she has the confidence of someone who knows what it is to be truly, unconditionally loved. She is in fact the happiest, kindest, most loving and giving person I know. Without a doubt. Does she get credit for her transformation? Of course not. It was all me. (*Kidding*, of course.)

Seriously though. Her default state is pure joy. She wakes up and greets the day out loud, with ebullient happiness just bursting out of her. *Good morning, kitchen! Hello, coffee maker!* It is absolutely adorable. I wish all of you could know how truly wonderful it is to spend your life with someone who constantly lifts up your spirits with her infectious happiness. Someone who knows how to laugh at life and not let the little things bring her down most of the time. Someone who would do an actual dance every time I came home, because she couldn't contain her happiness to see me.

Here's something I find quite funny. In this new life's mission that I've undertaken, I want all the people in my life to think of me (and remember me later) as the kindest person in their lives. And I honestly think I've made some real progress towards that goal with quite a few of my lifelong friends. Yet, for Julie's family, it would never even *occur* to them to think of me as the best person they know in that regard. That's because whenever they see me, I'm always standing next to Julie. And her joy shines so much brighter than mine ever will. I would have to up my game by *orders of magnitude* just to play on her level.

Trust me. I'm not saying this because I love her—which I do, implicitly. I'm saying this because it's true. It's like she walks the earth with beams of light just radiating out of her. I joke with her about the sunbeams shining out of her ass all the time. As we've traveled a small part of the world in these later years, I've seen how people are drawn to her and her happiness. Nothing makes me happier than seeing that and thinking about how we both used to be in comparison, as immature kids.

OK. Have I turned all of you off completely? Do you think I'm a moonbeam wacko? That's OK if you do. The Jeff from 30 years ago would have rolled his eyes at all this and probably thought the same thing. But am I getting through *just a little*? Is there a little voice in the back of your head saying, *you know, it would be nice if this actually worked*? Well, why not just take a *tiny* sip of the Kool-Aid with me and align your goal of personal fulfillment with mine?

Strive to be a happier person. Endeavor to spread a positive attitude to those around you. Work towards making the world a better place by not giving negativity a place to grow and fester. Pure happiness and optimism don't have to be relics of a distant forgotten childhood. You've heard the saying… Be the change you want to see in the world. I agree. *Live* the way you want the world around you to be. *Be* the change. Change the world for the better, simply by being better *yourself*. Be bold (as Goethe said) and mighty forces will come to your aid. Be bold with your happiness. I know it sounds corny and it takes a bit of bravery to get started. But I'm here to encourage you to take that first intrepid step. Do the courageous thing. Put that first bit of positivity out into your world. Then do it again, and again, and again. Eventually, that light will find a reflective surface to bounce off of and propagate. Together we can make this world a brighter, better place.

The dawn was breaking, a new day was rising, shadows creepin' over the fields
He watched the pink light, steal across the horizon, he realized, he saw with new eyes.
- Lyrics by Steve Hogarth

Every once in a while, you hear a profound truth that just stops you dead in your tracks. I am not a religious person. Long before it was socially acceptable to admit it, I was a confirmed atheist. An agnostic doubts; an atheist is convinced. I believe that religion on the whole holds back the human race from achieving its greatest potential, despite the very good intentions of many of its adherents and founders. And so, for decades, if someone asked me about religion, I would irrefutably state, "I'm an atheist. I don't believe in God."

Then I heard a story about the Dalai Lama. I'm paraphrasing here, but I heard that someone asked him what the one true religion was.

He smiled in that playful way he has and said, "There is only one true religion."

He pointed at the asker of the question and continued, "It's not yours."

He then pointed to another person and said, "And it's not yours."

"The only true religion... is *kindness*."

Never have truer words been spoken. Now that's a guy I could get behind if I had to.

So let's bring this full circle now. I've told you how I've made major strides in becoming a happier, less negative person. I've told you how I wished that I could have changed the world on a macro scale, but am succeeding (hopefully) in changing it bit by bit on a *micro* scale. This realization has brought me great contentment in my later years.

And then I had that night in Durham, North Carolina. A night when I was told that (even though I didn't find the cure for cancer or some other huge discovery), maybe I did in fact change the world. ...For this band, and for many others that followed. I was told that I altered the course of quite a few lives by being (as I stated at the very beginning of this book) the guy who was lucky enough to say the *right thing* at the *right time* in the *right place.*

Is it true? I can't really say.

I'm just glad to sit back now and think about the *possibility* that it's true... and smile.

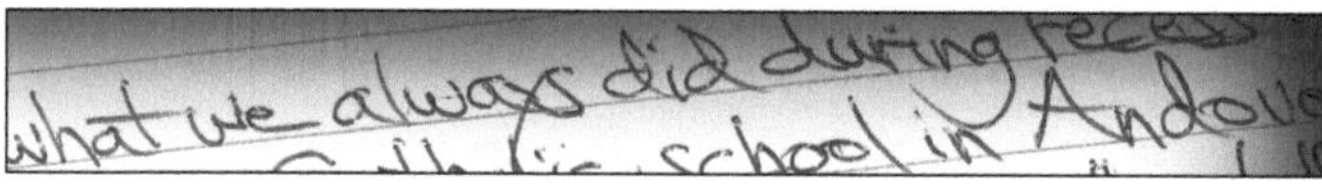

Story 18

I've saved this story for last. Too many times I haven't been satisfied with the result of the words I've strung together. And while I don't make any claim that this is great writing, I will say this. This one is true. It's the one time I feel I really hit the mark, to capture that moment exactly as it's remembered. It's also a perfect little window into what quintessentially makes me… me.

I don't remember the actual names of the kids in my first-grade class at St. Augustine's. (I transferred to the Andover public school system for second grade after the divorce.) But I do remember their faces and can picture them clearly, even today. So I've assigned them names for the purpose of the narrative. Other than that, all I have to say is… if there's one story I hope you enjoy, it's this one.

These are the days that will flash before our eyes at the end
These are the moments burned into the sacred places of our hearts.

- Lyrics by Steve Hogarth

Age 6, Grade 1, 1973
"Chasing Timmy"

It's what we always did during recess at St. Augustine's Catholic school in Andover—chase Timmy. Timmy Bouchette was the tallest kid in first grade and by far the fastest of all of us. During recess, all the boys would make a game of seeing who could catch Timmy first. If anyone else tried to be the runner, it never lasted as long. So, we always chased Timmy.

I had only been at this school for a year but I had already made some solid friends. First and foremost among them was Eric Carter. Eric always got a lot of attention from the nuns because for as long as I knew him, he had thick bandages beneath a white patch on his left eye. Over this monstrosity, his black-rimmed glasses stretched across his face to bring some clarity to his distorted world. Eric had straw-like blond hair and was as short as I was. So, of course, I liked him. Eric wasn't that fast but still participated in the chasing game, to the chagrin of the nuns. His left eye was apparently still functional under all those bandages since he didn't suffer from a noticeable lack of depth perception.

Another friend was Matt Larson. Matt was the token fat kid. He had straight, light brown hair that lay flat on his head. He was a full two inches taller than I was (making him seem a mountainous boy). Matt soon learned that not only would he never catch Timmy, but he looked absurd in the attempt. So, he'd sit on the sidelines and watch the game (sometimes chewing on a smuggled candy bar). To round out my collection of friends, there were also Bobby and David.

There were other boys that played the game, and some that didn't. Then, over on the other side of the playground, there were the girls—jumping rope and playing hopscotch and frolicking about in their more-subdued way. From this distance, they were an indistinguishable mesh of plaid green skirts—the female school uniform.

This day was a day like most others, except for the fact that I was feeling something I couldn't quite understand. It was like a monster of anxiety had been creeping up on me and suddenly had me surrounded before I knew what was happening. So, what did I do? I did what most six-year-olds will do when faced with a monster. I ran. And what did the rest of the first graders do? They chased me of course.

I ran with all the speed my little legs could pump out. And as I ran, I railed against the mental monster.

Why did Mom and Dad fight all the time?

And why did Dad have to move away?

And why does my brother always hit me and be such a jerk all the time?

And why do the nuns whack their rulers on kids' knuckles just for *talking*?

And why did Sister Mary smack Eric in the face and scream at him just because he was scratching under his eye patch?

And why do I have to go to this stupid Catholic school when Jessica goes to the public school?

And why do the "Big Kids" always steal our kickball or wreck whatever game we're playing in the street?

And why is everyone so *mean* all the time?

I thought I'd run away and never be caught. But sure enough, I soon felt Timmy's hand on my shoulder, as he was the first to catch up with me. Out of breath, we came to a stop and then he saw my eyes and the water that had collected there.

"What's wrong?" he asked with the parental concern of a leader.

Eric, Bobby and all the others showed up next (even Matt who must have sensed that something important was happening). There I was, exposed before my peers. Eric asked what was wrong. Matt did too. But it was Tim I answered.

I looked him straight in the eye.

"Nothing's wrong," I lied. "I just wanted everyone to chase me for a change."

Timmy knew it wasn't true, but he also saw the look in my eye that begged him to let it go. A few more unspoken words passed between us before he reached his decision.

"All right, everyone," he commanded. "Today we're chasing Jeff." And he turned back to me.

"Well... let's go," he motioned.

I felt a wave of relief wash over me.

And off I went.

And off they came after me.

And as I ran, I realized that today Timmy wasn't running as fast as he could have, so the game would last a little longer.

And as I ran, I realized that if you had friends, then things weren't really that bad.

And as I ran, I felt some of the demons fall behind as they were left in my wake.

And as I ran, my eyes began to dry.

And as I ran, I felt the sun in the clear blue sky spotlighting me in this moment of clarity.

And as I ran, a smile slowly crept across my face.

Chapter 39

That's All, Folks

Well, I guess that's it. I really hope you've enjoyed this time we've spent together. If you have, I'd love to hear from you. Feel free to drop me a line sometime. You won't find me on social media. (To quote Bill Belichick, "*I don't do any of that SnapFace.*") My non-work e-mail address hasn't changed in 20 years though. If you know it from back then, you know it now. If not, you can write to me at marillionjeff@gmail.com — the address I created for everything to do with the book.

If you didn't like the book and you want to tell me I suck as a writer, you can feel free to do that as well. I won't be insulted. Well, unless insulting me will be your goal. Then tell me beforehand that that's what you're going for, and I promise that I will in fact be insulted by your insult. It's the least I can do if you've read this far.

There are still a few pages left to turn through before you reach the back cover. I've decided to include some of my family's recipes and some that have special meaning to us. Give them a try if you're up for it. I hope you'll find the waffles and crepes recipes worth whatever you've invested in this (be that time or money or both).

I think it's quite intriguing how this story all unfolded. For a while there, I was quite well-known in certain circles. Yet after a number of years had passed, it seemed like no one would remember what we had accomplished or the role I had played in it. Then in the past 5 years or so, it seemed like there was a resurgence of interest in the Tour Fund and how it came to be. Suddenly there were interviews and articles all over the place mentioning me by name as the creator of the first online crowdfunding campaign.

I had a lot of fun writing this book. It was a lot of work too—don't have any doubt about that. But reliving these moments has been a life-affirming experience for me. So, thanks for coming along with me on this ride. I hope you enjoyed it to even a fraction of the degree that I did. Again… thank you.

Jeff Pelletier

December 2021

Chapter 40

An Unanticipated Epilogue

I lied. That's not "all," folks. Sorry. It's now close to the end of 2022—a year that saw the return of the Marillion conventions in glorious fashion. Julie and I attended the one in Montreal, Quebec as we always have. I expected to be blown away by the new album, particularly the epic closing track—"Care." And I was. The "Angels on Earth" closing section soared to transcendent heights. Heights that were further heightened in context by the long-awaited return to live music (after 2 years of yearning).

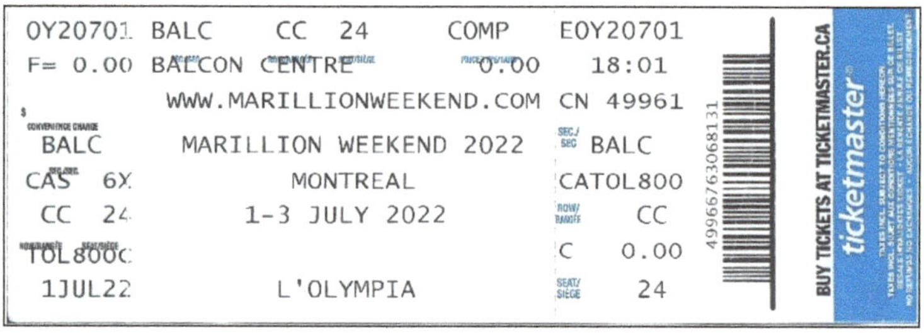

What I *wasn't* prepared for was how "The Crow and the Nightingale" would be the highlight of the *An Hour Before It's Dark* night–with a guitar solo that was soaring, exhilarating and cheer-inducing. It took center stage instead of being almost overshadowed by the amazing Choir Noir—as it is on the album. Well done, Steve. And well done, Marillion. As always.

How do you keep tapping into that figurative gold over and over? It's almost like you guys collectively create something beyond music. It's transcended that distinction to become something *more*—something akin to emotional telepathy. And I am just amazed by that.

We were backstage talking to Pete on the final night, and I summoned the courage to get over my inability to ask people for small favors. I mentioned that I wrote a book about the Tour Fund (with the intention of asking him for a brief quote like the one I got from h about the show dedication in North Carolina).

Pete responded jovially, "You wrote a book? That's great." Then before I could ask, without any prompting from me, he said, "You know. What you did wasn't just for everyone here. You did it for the entire world."

He then continued with another amazingly quote-worthy statement of praise. I laughed and said, "That right there! That's perfect. Can you write that down and send it to me?" He, of course, happily agreed and gave me his address, since he's a great guy.

Exchanging e-mails later, we both had forgotten exactly what he had said, but he did himself one better. He provided the following quote which was so good that if you close this book, you'll see it in part on the front cover.

A Note From Pete Trewavas

It's not often that someone can claim they helped shape the future. But Jeff is one of those rare people. This all happened way back in the infancy of the internet. And Jeff (with the enthusiasm of youth, the faith in his fellow man and the generosity of an out-and-out fan) decided he wanted to help us, "Marillion," his favourite band, tour America.

As I'm writing this, I can see how ridiculous it may sound. However, he turned his wishes into a dream come true, in this quite remarkable true story.

How do I know this? I was there.

Pete Trewavas
November 2022

You can't ask for much more than that can you? And how do you properly thank someone who gives you such a fantastic compliment? I honestly don't know. All I can do is say, "Thanks, Pete!"

And thanks to you readers. You have reached the "end" (finally) of the main section of the book. (What follows is all supplemental material.)

Enjoy being alive. Enjoy the music. And be kind to each other. See you at the next show!

Unloading our cargo, the thrills of the night
Our boxes of noises, our boxes of light
We are parties that travel, movies that move
Jokes that span fields, conversations that chatter
From one town to another, laughter that passes thoughts on the wind.

- Lyrics by Steve Hogarth.

Recipes

Preface

My dad was a fantastic cook. He could make a great tomato sauce, an incredibly hearty & tasty beef stew, the best rotisserie chicken you've ever had, and so many more delectable meals… But for me, it all came down to his breakfasts. Some of my greatest memories of my teenage years (when my brother and I were living with my Dad), were lying in bed on weekend mornings with those fantastic breakfast aromas drifting into my room, beckoning me to rise. I was usually so lazy that I would continue to lie there just enjoying the smells until the knock eventually came on my door saying it was time to either get up or miss out. (Oh, how I miss those days of being able to sleep for over 12 hours straight…)

Dad would usually rotate the main breakfast course from week to week. One day it would be his French crepes, the next waffles, the next French toast, and sometimes it would be omelet day. And with each of those days, the main breakfast entrée would be accompanied by bacon, sausage, his famous home fries, grilled English muffins and scrambled eggs (the eggs on every day but omelet day, as you might guess).

Have you ever had waffles that are light and airy on the inside with just a hint of crispiness on the outside? Waffles that are so good, you don't have any of that heaviness in the stomach after you've eaten one and can't help but reach for another?

In his later years, my Dad spent an astounding amount of time and effort crafting the perfect waffle recipe. In my memories, I picture him still in his kitchen, searching for and eventually creating that elusive perfect batter like an alchemist.

Do you want to try to recreate what he considered his final and best culinary achievement? Is experiencing the *best waffles ever* enough motivation for you to commit *the night beforehand* to having them for breakfast the next morning? If so, are you willing to follow a recipe to the letter? It's not complicated, but the times and temperatures and separate steps are important.

Well, if you've answered in the affirmative to all of those questions, then here is my Dad's final culinary achievement—The Greatest Waffles you'll ever make….

Some points of clarification from the master's son (that's me for those of you not so familial-ly inclined): First, don't buy the quick-rise yeast. Get the normal yeast packet. And if you're like me, you'll probably let those other packs of yeast sit in the kitchen cabinet for like a year. If you do, you can still use them, but the lightness and airiness of the resulting waffles will be *significantly* diminished. So, use *fresh* yeast if you want them to win that blue ribbon. Again—*fresh*, and not quick-rise yeast. And you want to use real dairy milk. We've tried it with vanilla almond milk and they still come out fine, but the batter is a lot gooier and the waffles don't rise to give you that great light airiness on the inside—which is the whole point.

Next, technique: Don't completely fill the waffle iron with batter. Ideally, the edges of the iron should be unfilled to give the batter room to expand without popping the iron open mid-cook. Waffle irons may vary. Dad bought me a nice *Krups* waffle iron as a Christmas present one year. It's got a cooking surface about 6" x 10" in size. I recommend it. It's small enough to fit in a drawer and has a quality non-stick surface, letting the waffles pop out easily.

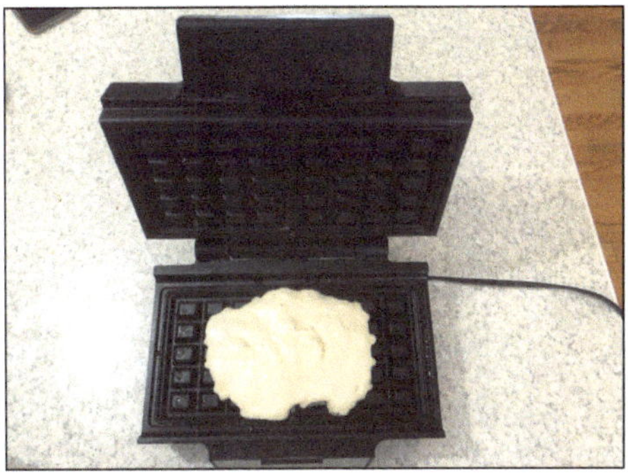

I find that I have to cook the waffles about five minutes (not four), if I want a little crispiness on the surface. Run the iron through one heating cycle empty before you start. And yet, the first waffle will still probably take a minute longer than the rest. I can't explain that. You'll find the right timing for your iron as you experiment with your cooking implement and become a waffle master yourself.

And one final thing. I know it's expensive, but if you can afford it, real maple syrup is vastly superior to the artificial fructose syrup simulations. My dad used to put some water in a shallow pan on very low heat and let the little plastic jug of real maple syrup warm in the pan. Not necessary, but it's a very nice flourish to have warm genuine syrup to drizzle upon those perfect waffles. Enjoy!

Recipe 1

Dad's Greatest Waffles Ever

To serve at breakfast, *start them the night before.*

Ingredients

1/2 cup lukewarm (about 110° F/43° C) water
1 package yeast
1 teaspoon brown sugar
1½ cups lukewarm dairy milk
1/2 cup lukewarm melted butter
1 teaspoon salt
2 cups sifted all-purpose flour
2 eggs
Pinch of baking soda
2 teaspoons vanilla extract

Batter after it's bubbled up overnight (above) producing air pockets inside the finished product (below) for that perfect waffle.

Steps

1. In a large mixing bowl, add the water, yeast and brown sugar. Gently stir.
2. Let stand 5 minutes, and then add the milk, melted butter and salt. Beat in the sifted flour.
3. Cover the bowl. Let stand overnight (or at least 8 hours) at room temperature.
4. When it's time to cook, separate the yolks and whites of 2 eggs. Beat the whites until stiff. Beat the yolks separately.
5. Fold in the yolks and the beaten egg whites and add the baking soda and vanilla extract.
6. Cook in pre-heated waffle iron to taste, usually about 4 minutes.

Makes 6 to 8 "great" waffles. Bon appetit.

Recipe 2

Dad's Regular Waffles

Let's face it. Sometimes you want waffles and you can't travel back in time to make the waffle batter the night before. Accordingly, in addition to perfecting his ultimate waffle recipe, Dad also had his backup plan. If you need to make waffles right away, here's a good option for you. They won't be as light and airy, but they won't be heavy and disappointing, either. Note that we use canola oil instead of corn oil.

Ingredients

- 2¼ cups all-purpose flour
- 4 teaspoons baking powder
- 1/2 teaspoon salt
- 1½ tablespoons brown sugar
- 2 eggs
- 2¼ cups dairy milk
- 2 teaspoons vanilla extract
- 1/2 cup corn oil, canola oil or butter

Steps

1. Sift together the all-purpose flour, baking powder, salt and brown sugar.
2. Separately, beat the eggs, then add the milk, vanilla extract and oil (or butter).
3. Add the liquid ingredients to the to dry ingredients, beating only 'til moistened.
4. Bake in pre-heated waffle iron (about 4 minutes).

A double batch of perfect waffles.

Recipe 3

Dad's French Crepes

Crepes were a part of my childhood for as long as I can remember. Dad's side of the family was French, and crepes were part of the heritage. They were such a superior alternative to the standard pancake that I was surprised later in life to learn how easy they are to make. I was intimidated the first time I tried to make them on my own, but all that worry was unnecessary. Give it a shot and they'll turn out wonderful. Trust me.

The one thing you need to know is that the amount of oil you use can modify the texture of the crepes to a great degree. My grand-mama, Leonida Pelletier (affectionately called "Gummama" by her many grandchildren), made her crepes with more oil. This resulted in a crispier crepe. My Dad used less oil for a softer more delicate crepe. I absolutely prefer Dad's version. I use about a teaspoon of oil in the pan for the first crepe, and then about half of that for each crepe after the first one.

You can use almond milk instead of dairy milk, since you don't need the crepes to rise like waffles or pancakes. Again, I usually use canola oil instead of corn oil.

For our weekend breakfasts, Dad used to double the recipe. As you know, you can fill a crepe with just about anything. Fresh fruit and whipped cream are standard. Nutella is a great option for chocolate lovers. In his later years though, Dad would always fall back on the very simple option of rolling them up with just a bit of brown sugar sprinkled inside. Once again, real maple syrup is the perfect accompaniment to this simple perfection.

The Pelletier Crepe Recipe

Ingredients

3 eggs
1¼ cups milk
3/4 cups all purpose flour
1 tablespoon brown sugar
1 teaspoon vanilla extract
1/2 teaspoon salt and/or a dash of cinnamon (Optional)

Dad used a blender to mix the ingredients. You can do it easily with a whisk though if you don't want to break out your blender.

Steps

1. Blend the eggs until thick and lemon-colored.
2. With blender on low, slowly pour in milk and flour.
3. Add sugar, salt & vanilla extract.
4. Blend till smooth.
5. Pour some oil in a moderately hot skillet (more oil for more crispiness).

6. Spread the batter evenly in the pan to make thin crepes (see tip below). Turn when golden brown. Add additional oil for each crepe.
7. Apply brown sugar and/or whipped cream, strawberries, blueberries etc. and roll up crepe. Add syrup to taste.

Pour in one ladle full of batter. Pick up the pan, tilt it slightly and rotate through 360 degrees to spread the batter out and fill the pan (left). The first crepe will take a couple of minutes to cook before you flip it. After that, you can flip them after less than a minute per side—once you see some bubbles (left) for a perfectly cooked crepe (right).

One batch of batter makes six perfect crepes (left) ready to roll up (right). This time we rolled them up with fresh blueberries & banana slices inside and blueberry syrup & whipped cream on top (below). Pure heaven!

Recipe 4

More Breakfast Tidbits

Dad's Home Fries

Potatoes have gotten a pretty bad rap as of late. I'm guilty as well of falling for all of the "*Avoid Starch at All Costs!*" hype. (Hey. Remember when they told us all to stop using butter and use margarine instead? Then we later found out that margarine was *WAY* worse for us?) So, if you want to enjoy some good breakfast potatoes, then do it. Just about everything is fine in moderation. Here's a tip that many of you probably already knew, but I didn't. Heat up the oil in the pan *before* you put in the potatoes. If you don't do this, more of the oil will soak into the potatoes and they'll be over-soggy. Heat the oil *first*, and a nice barrier of crispiness will prevent soggy potato syndrome from prevailing.

I also recommend dicing them into small pieces—less than half an inch if you have the patience. I take the potato and cut it into discs and then dice up the discs (pictured below, at left). In my opinion the higher the surface-area to potato-mass ratio, the better. Larger pieces will result in more of that bland mashed potato taste at the core of each home fry. Smaller pieces yield more of that tasty, fried goodness.

You don't need to submerge the potatoes in oil. Just use enough oil to cover the surface of the pan. Make sure you start the potatoes first when you're making other breakfast dishes along with them, as they'll take the longest to cook to the desired crispiness. Stir them every few minutes and season as desired. I used some black pepper and a paprika-based barbecue rub seasoning for the potatoes pictured below (right).

Baking Bacon

Bacon is easy to fry but time-consuming—especially since you usually can't fit the whole package in one pan at a time. Did you know you can bake it and it comes out great without getting your stovetop all splattered with grease? I didn't. High heat works well. Bake at 425 degrees for 15 minutes pre-heated, 20 to 25 minutes if you don't wait for the oven to reach temp. This is a good alternative to frying (since I've read that microwaving bacon can produce some unhealthy compounds).

Use a baking sheet with *raised* edges—otherwise, the grease will run off into your oven. *Bad.* I use parchment paper or a piece of foil with the edges raised and corners folded to try to contain all the grease. You'll end up having to wash the baking sheet anyway, but nothing gets baked onto it, at least. You can overlap the bacon strips a bit to fit them all on one sheet like below.

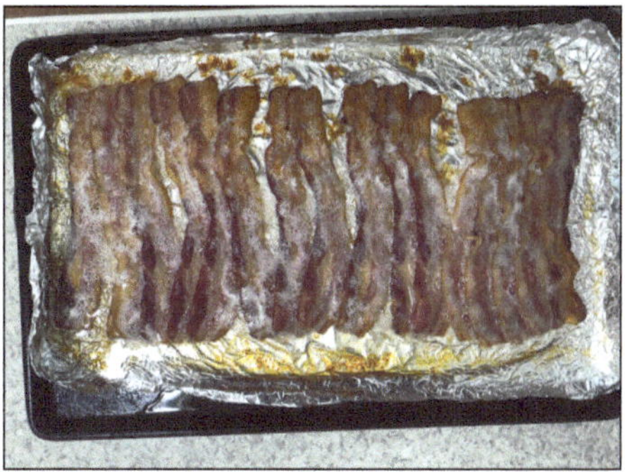

When you're done, don't throw away all that bacon grease. Save an empty metal can the next time you open a can of veggies or soup. Pour that bacon grease into the can and throw it in the fridge. It will solidify to an opaque, gray-white, paste-like consistency (which will turn perfectly transparent when melted later). The next time you make eggs, try using a bit of bacon grease in the pan instead of butter or oil. It gives the eggs a nice bit of crispiness and a hint of bacon-y flavor.

Eggs in a Minute

I make eggs all the time. And one thing I hate is washing the pan afterward. I never considered making eggs in the microwave until I saw an article about how good the results are if you own a ceramic ramekin dish. What you do is this. Add about a tablespoon of water to a ramekin. Then just crack an egg on top and season it if you want (i.e. pepper and/or salt). Put it in the microwave and cover it with an empty dish. Blast it with radiation for about 45 seconds and the water will help give the egg that perfect poached quality.

Note how an extra 10 seconds or so will change the yolk from "over easy" consistency (top) to "over medium" (bottom) or unfortunately (in my opinion) "over hard."

Don't be alarmed if the dish rattles because some of the steam needs to escape before it finishes. Also, since microwaves vary in power, try starting at 40 seconds. Then go 10 seconds more at a time until you get the yolk consistency you like. As I said, mine usually comes out perfectly cooked with a still-runny yolk at 45 seconds, though it can vary with the amount of water I use.

Use a fork or spoon to hold the egg in place while you drain the excess water. Then the egg slides out perfectly. No pan to wash. The ramekin usually just needs an easy rinse to clean it as opposed to scrubbing a pan. And the added benefit of the ramekin is that the egg will be perfectly shaped to slide right onto an English muffin for the ideal egg sandwich. Add cheese, ham, bacon, tomato, avocado and/or whatever you like or have on hand.

Liquid Egg Whites

Speaking of eggs… My friend Mark Perlberg recently told me how when he makes an omelet, he often uses four eggs but throws out two of the yolks. I used to do the same thing (with three eggs minus one yolk). Remember the *No Fat* craze? Now they're saying the nutrients in the yolk are beneficial enough to justify the fat. I was never a full-on "egg-white omelet" guy, but I still like to try to strike a reasonable nutritional balance between fat and protein.

A few years back I discovered that you can buy a pint of liquid egg whites. Now don't confuse what I'm talking about with that artificial Egg Beaters crap. I mean *pure* 100% liquid egg whites. Here in New England, you can even buy them in an economy store brand (*Market Basket*). So, when I make an omelet, I usually use two eggs (including the yolks). Then I pour in another egg or two's worth of egg whites from the carton. Perfect fat/protein compromise and no yolk waste. I hate wasting perfectly good food.

French Baguette Toast

For some reason that didn't seem strange to me at the time, whenever we ordered Chinese food here in New England, it came with a big baguette of French bread. If you had a hefty takeout order (as we often did), you'd get two baguettes. A large portion of this bread would often go uneaten. So my Dad would slice it into 1-inch pieces and then put them in the freezer in a Ziploc bag.

The next time he would make French toast, out came the "Chinese-food" bread from the freezer. And tasty French toast was soon to follow. Also, when you've got a chunk of leftover Italian bread that's starting to go stale, you can do the same. Cut it into thin slices while it's still hard, then microwave the slices just enough to soften them a bit. Now they'll soak up the egg mixture and make a great breakfast. Even that tiny end piece—as long as you shave off the crust from the very end.

Couldn't resist making a few for visual documentation – French bread (left) and Italian bread (right) down to the small end piece.

Recipe 5

Mom's Tomato Sauce

When it came to food, I was a pretty lucky kid. I grew up in two different households with two excellent cooks. Mom was Italian and was a master of the cuisine. I never appreciated it enough back then. Instead, I just pouted incessantly whenever she made one of Alan's favorite meals—like Shepherd's Pie. *Blech!*

Tomato sauce was always a staple in our house. It's easy to make. Seriously. Time does all the work for you. Just gather these few ingredients and combine them as instructed. Let them simmer in that big pot all afternoon and the result is a great sauce to go with so many meals. I often add a tablespoon or two of Italian seasoning, instead of just relying on oregano to do all the heavy lifting on its own. Also, if the cans of tomatoes don't have basil in them, add some yourself if you have it on hand. And if you want low sodium, you can skip the salt. You won't miss it.

The first time you make the sauce, make the full recipe. Throw a good amount of it in the freezer if you need to. It's only *slightly* better when it hasn't been frozen, so you won't see a significant drop-off in quality after you thaw the frozen portion. And you'll be happy after a few weeks go by and you're ready for *Italian* to re-join the dinner rotation once again. Then just pull that container out of the freezer.

The environmentalist in me hates buying those little 6 oz. cans of paste. You can buy 12 oz. cans, and sometimes you can even find paste in the 18 oz. can. With the 12 oz. can, you could just make two-thirds of a batch using only two big cans of tomatoes. Or use one 12 and one 6 for a full batch.

My mom insists on quality name-brand tomatoes. She prefers Cento or Pastene. Honestly, I've found that they all make great sauce. I buy whatever's on sale. (*Don't tell her.* We've even used Market Basket store-brand canned tomatoes.) Mom uses peeled tomatoes, and those large tomatoes *will* break up as they simmer and get stirred. We often get the crushed tomatoes instead though. And sometimes for the third can, we mix in a can of diced tomatoes. You can try different combinations to land on your favorite consistency.

Also, let me add this final note. I do *not* like Ragu or most of the jarred sauces. They just taste artificial to me. If you live in the States, though, we find that Trader Joe's jarred marinara sauce is great to have on hand for a quick acceptable sauce in a pinch. Now, with that being said, let's get our sauce on…

Tomato Sauce

Ingredients

3 28 oz. cans Cento Italian peeled tomatoes
3 6 oz. cans of Contadina tomato paste
1 tablespoon of oregano
2 teaspoons of fresh ground pepper
1 teaspoon of salt
Water, (3 tomato **paste** cans—18 oz. in all)

Steps

1. Put tomatoes in large saucepan on low heat.
2. Add tomato paste. Fill each tomato paste can with water and add to saucepan.
3. Add oregano, pepper, and salt.
4. Add meatballs (see next recipe).
5. Let simmer on a low burner setting for three to four hours with cover partially off the saucepan so the steam can escape. Stir occasionally so it won't stick to the bottom of the pan.

Add all the ingredients in a large pot (left). Stir it all up, then add in the meatballs and sausage from the next recipe (right). Let it simmer for several hours (with occasional stirring) and you'll be ravenous from the aroma once it's time to eat.

Recipe 6

Mom's Meatballs

You've made the sauce, and now it's time to make some meatballs. Now I'm going to tell you something that's going to sound crazy, but you have to trust me on this. Are we cool? Will you just agree to go with it? Even if it sounds nuts? All right. Remember, you *promised.* OK. Here we go. Our family puts raisins in their meatballs. *What!* Why on earth would you do that? Whoa, whoa, whoa... Settle down and let me explain. And remember, we had a *deal...*

The raisins (when cooked in the meatballs) give the meat just a hint of sweetness. They will plump up inside the meatballs (even if they're completely dry) and then infuse the meat as they release their raisin-y goodness. While they won't disintegrate, they'll end up with the same consistency as their surroundings. So as you eat the meatballs, you'll barely notice the raisins are there.

OK. You *promised.* So, don't chicken out on me now. You're going to try this at least once with the raisins. Here's the other thing. If you want to do this right, *you have to make tomato sauce.* The most important step in making great meatballs is to drop them in the sauce and let them soak in that flavor for a few hours. This also helps in the fact that you don't have to overcook the meatballs before you add them to the sauce. Just brown all of that surface area and then let the sauce bring it home.

When buying ground chuck, we usually get the 85% lean. If you go with the 90 or 95% lean meat, you're actually paying more while the *Law of Diminishing Returns* works against you. Why is that? Because when you fry leaner meat in the pan, less fat or grease comes out. Consequently, the final result ends up being not that much different than with the 85%. So, we only buy the super lean meat if it's on sale.

We also usually get a package of Italian sausage to make at the same time. We brown them in the same pot the sauce will go in to add a bit more flavor to the sauce. Then slice them up and add them to the pot when you add the meatballs.

Meatballs

Ingredients

1 pound ground chuck
3 eggs
1 cup plain breadcrumbs
1 teaspoon fresh ground pepper
1/2 teaspoon salt
1 tablespoon parsley
1 tablespoon grated parmesan cheese
1/2 teaspoon minced garlic
2 tablespoons raisins
Extra Virgin Olive Oil

Steps

1. Mix all ingredients together (except for olive oil) and form meatballs about 1 inch in diameter.
2. Fry in olive oil on all sides until brown.
3. Add meatballs and olive oil to Tomato Sauce.

Add the ingredients to a large mixing bowl (left), then get your hands in there to mix it all together. Form meatballs (right) to your preferred size.

Again, brown the outer surfaces only in a pan with just a little oil (you'll get more grease in the pan, the more you cook). Then add them to the sauce as you go (right).

Recipe 7

Mom's Lasagna

So now we have the sauce and we have the meatballs. Should we go for the trifecta of a perfect Italian meal and make a lasagna? Why not? Lasagna is also incredibly simple. Mom will tell you to only use a fancy pasta brand like DeCecco (her current favorite). I can tell you that it's going to come out great no matter what lasagna you buy. (Just don't get that *bake-able* lasagna. It comes out OK, but it is *vastly* inferior to the pasta you have to boil before you assemble.)

Lasagna all comes down to the sauce. Use the homemade sauce we made a couple of pages ago or a quality substitute of store-bought sauce like the Trader Joe's marinara we talked about before. (No Ragu or any of those salty stand-ins.) Next: We always get the *part skim* mozzarella. No need to go with the whole milk. You can't tell the difference.

Lastly, I have some work friends who insist that it's not lasagna if it doesn't have ricotta cheese. Well, they're just wrong. Sure, you can add in your "filler cheese" if you want. But then you miss out on the simple perfection of the dense pasta/mozzarella/sauce construction. Furthermore, when I ordered a small serving of "lasagnetta" in the Duomo square in Milan, I got the most amazing fresh spinach pasta layered (off-center in a lovely pattern) with just mozzarella and sauce. Best pasta dish I've ever had. *In my life.* By *far*. So, I guess they're wrong about lasagna in Milan, Jim? I don't *think* so!

Cut the mozzarella into thin slices, not chunks. It melts and spreads out better that way.

After you boil the pasta, separate the noodles to cool before they do that Vulcan mind-meld to each other. Then set aside four strips (without any defects) to use as the top layer when you assemble. Mom puts grated cheese on top of each layer of sauce. I just use a little on the top. I want the mozzarella to be the dominant cheese taste.

OK. Time to assemble. Note that you can assemble the lasagna and keep it in the fridge for a couple of days before you bake it. Handy to do if you want to save time on the night you'll be serving it.

Lasagna

Ingredients

1 lb. box lasagna (DeCecco or Ronzoni Brand)
1 recipe tomato sauce (previous recipe)
1 pound whole milk Sorrento mozzarella cheese, sliced thin
Grated Romano or Parmesan cheese

Steps

1. Boil lasagna in water as directed on box.
2. Drain in colander and separate the lasagna pieces.
3. In a 9" x 13" rectangular pan, cover bottom with tomato sauce.
4. Assemble layers with lasagna, then sliced mozzarella cheese, then tomato sauce, then grated cheese.
5. Continue to make layers until all the lasagna and cheese is gone, ending with sauce on top.
6. Cover with foil and bake in 400° F (204° C) oven for about 30 minutes.

Fully assembled lasagna, ready to bake. It's good to try to have one or two extra slices of mozzarella to break up and spread on top (with a little grated parmesan) when you're done layering.

A freshly-baked serving of lasagna (below) with sauce and meatballs—melted cheese deliciously oozing out the side...

Recipe 8

Veal Parmigiana (or Chicken or Eggplant)

Do you have a problem with veal? It's OK if you do. If so, you can substitute chicken cutlets or eggplant (sliced thinly) in this recipe. All three come out great. If you want to serve a big Italian meal, adding one of these options to the lasagna with sauce and meatballs is an excellent choice. An additional benefit is the fact that you can bake the cutlets and the lasagna at the same time at the same temp. So, it makes it that much easier to serve this meal if you're entertaining while you're cooking.

Not much to add with this recipe. Just don't forget to heat up the oil before you add the cutlets. And if you decide to make several batches of these, you should use more than one baking dish. If you stack the meat in layers in a single dish, they won't fully cook in the same amount of time. We made that mistake once with several layers of stacked chicken cutlets, and the ones in the middle were still semi-raw in the middle when we served them. Yikes! (*Merry Christmas, Family. Salmonella is your bonus gift this year!*)

Veal Parmigiana

Ingredients

1 pound veal cutlets or whole veal loin cut into thin pieces (1/8" thick)
3 eggs, beaten
Plain bread crumbs
Salt, pepper and garlic powder
Parsley flakes
Extra virgin olive oil
Mozzarella cheese
Tomato sauce (see Recipe No. 5)

Steps

1. Put olive oil into frypan and set heat to medium.
2. Mix breadcrumbs, salt, pepper, parsley flakes and garlic powder and put into bowl.
3. Beat eggs in a separate bowl.
4. Dip veal cutlets in eggs and then into breadcrumb mixture.
5. Put into hot olive oil in frypan and let cook until crispy, then turn over and let other side cook until crispy (total of about 20 minutes).
6. Put all the cooked cutlets into a baking dish; cover with Tomato Sauce and mozzarella cheese and bake at 375°F (190°C) for about 20 minutes (until cheese is melted).

Fry up those breaded cutlets in a bit of oil (above). You can serve them as is or put a little sauce in the bottom of the baking dish (below) if you want to "parm" it up.

Cover with sauce and shredded mozzarella (left). Then bake as instructed for one of my favorite meals (right).

Recipe 9

Julie's Homemade Chicken Pot Pie

Julie is constantly saying she's not a great cook. I have no idea why she says this, because so many of the dishes she makes are fantastic. One of my favorites is her chicken pot pie. A homemade pot pie is just the *embodiment* of comfort food. So good, fresh out of the oven. And so good, reheated as leftovers. It's one of the few meals I don't get sick of when I have it three or four nights in a row. She'll make that filling on the stove top and it just has that perfect gravy-like consistency when the pie is assembled and baked.

A standard pot pie is made with peas and carrots, but I am not a fan of the pea. (Even if Forrest says, "*Jennie and me was like peas and carrots.*") So we've gotten into the habit of using an assortment of different veggies. We usually use a can of no-salt carrots, some Trader Joe's frozen white corn kernels, then some finely chopped broccoli and/or cauliflower, and even add some chopped zucchini or frozen spinach sometimes… Really, you can do it with any vegetables. And it's so much better than the standard peas and carrots version. My strong preference is that whatever veggies you use, chop them up into *small* pieces. I prefer not getting a bite of pie that's all one ingredient. I want that satisfying mélange of tastes. So, my rule of thumb is to chop up those veggies til you've got pieces the size of kernels of corn.

Another time saver is pre-cooked chicken. If you want to skip that step you can buy Perdue Short Cuts carved chicken breasts. It's real chicken. Also, don't feel bad about buying the pre-made crusts. They make a great pie and the convenience is so worth it. They're in the red box in the refrigerated section of the store, next to the crescent rolls in those little cardboard tubes that startle the crap out of you when you twist hard enough to break that vacuum seal and they pop like a shotgun blast. You'll need to let the crusts come to room temp before they'll roll out properly. At this point, I'll turn it over to Julie. It is her pot pie, after all.

Julie says ...

The key to a really good chicken pie.... fill it to the brim with every veggie you have on hand! Don't just stick with carrots and peas. Boring. Throw in corn, broccoli, cauliflower... whatever you have in the freezer. Here is another tip... if you make the filling ahead of time (even days ahead) make sure you heat it up in a saucepan before putting it in the pie crust and baking! I learned the hard way that pie will be cold inside if you don't. And no one finds a cold chicken pot pie comforting.

One other tip: If you don't want to make your own pie dough, use Pillsbury pie crust. HUGE time saver — and you can hardly tell the difference.

Chicken Pot Pie

Ingredients

1/3 cup butter
1/3 cup all-purpose flour
1/3 cup chopped onion
1/2 teaspoon each, salt and pepper
1 ¾ cups chicken broth
2/3 cup milk
3 cups cut-up cooked chicken or turkey
Frozen corn, carrots, peas, broccoli, spinach...any veggies you want
1 box (15 oz) refrigerated pie crusts (2), softened as directed on packaging.

Steps

1. Let pie crusts come to room temp.
2. In 2-quart saucepan, melt butter over medium heat. Then stir in flour, onion, salt and pepper. Cook, stirring constantly, until mixture is bubbly. Then remove from heat and stir in broth and milk.
3. Heat to boiling, stirring constantly. Boil and stir 1 minute.
4. Stir in cooked chicken and vegetables; remove from heat.
5. Pre-heat oven to 425°F. (218°C)
6. Unroll 1 pie crust and place it in an ungreased 9-inch (2-quart) round glass baking dish.
7. Pour your *warm* chicken/vegetables pie filling into crust-lined dish.

Cook the chicken first and cut into small chunks (left). Then mix up the filling in a pot (middle) as instructed. Then with the bottom crust in the glass baking dish, pour in the filling (right). Make sure it's warm. Heat the filling first if you prepared it beforehand and just took it out of the fridge.

Continued on next page

8. Unroll remaining pie crust, fold it gently and make 4 to 5 slits in the top, then place it on top of the pie.
9. Turn edges of pie crust under, crimping the edges together.
10. Cover loosely with foil and bake about 40 minutes, remove the foil for the final 10-15 minutes to get that golden brown crust.

Put the top crust over the filling and crimp the edges of the crusts together (left). Cut a few decorative slots in the crust for steam to escape. The earlier you pull the foil off during baking, the more you'll brown the crust. This one (center) came out great. My previous attempt was done without the foil and the crust was overdone (right) but still tasted great.

Recipe 10

Mom's Southern Chicken Stew

OK. All this one comes down to really, is that it's just another chicken and pasta dish. But even though it doesn't look like there's much to it, it's got something about it that I really grew to love when Mom picked it up after she moved down to Carolina. (And no, James, she didn't just go there *in her mind*.)

Note the directions about the constant stirring as you add the dough to the stew. You definitely need to do that. A couple times I've tried to make this meal a little healthier by adding some veggies to it. But I have yet to improve on this dish that way. In fact, I usually end up detracting from its appeal instead of enhancing it. So I'd say, if you want to be healthy with this main course, just have some veggies on the side.

And finally, a note of personal preference. I think this stew is much better when you do a really good job of cutting and shredding the chicken into small pieces. You don't want one bite to be just a big hunk of chicken. You want it all to blend together as much as it can. Enjoy!

Chicken Stew

Ingredients

- 3 quarts water
- 3 chicken breasts
- 2 stalks celery, cut in 3" pieces
- 1 small onion, quartered
- 2 cups flour
- 1 egg
- 1/2 teaspoon vegetable oil
- 1 tablespoon butter
- Salt and pepper and parsley, to taste

Steps

1. Boil chicken in water with celery and onion for about 1 hour. When done, remove the chicken, onion and celery from the water. Let chicken cool. Throw away celery (it was added to flavor the broth).

2. Combine flour, egg, oil, salt and pepper. Mix and add water until you get a dough-like consistency. Roll out the dough with a rolling pin until about 1/16" thick. Cut into rectangles about 1½" x 3." Add one piece at a time to boiling broth that chicken was cooked in, and keep stirring or they will stick to the bottom.

3. Cut and shred the chicken and onion into small pieces and add to broth. Add butter and parsley to taste. Cook on medium heat uncovered for about 1/2 hour, stirring frequently enough to avoid the dough pieces sticking to the bottom (especially toward the end when it thickens up).

Shred that chicken thoroughly after it's boiled and cut the onion pieces (left). Roll out the dough 'til it's nice and thin (center) using flour on the board and the rolling pin. Then cut into pieces (right). Don't worry about making them perfectly shaped.

Add the pieces in one at a time, constantly stirring (left). When you're done, add the shredded chicken and the rest of the ingredients (right). Keep stirring enough to prevent anything sticking to the bottom of the pan and burning.

Final result is a nice thick chicken stew.

Recipe 11

Helen's Thanksgiving Squash

Back in my Polaroid days, our division, PGI (*Polaroid Graphics Imaging*—which was later spun off to become *Latran Technologies, LLC*), used to give all its employees a frozen turkey as a nice little holiday bonus. This giant rock of a thing would sit in my freezer for several months and then each summer Julie and I would host a turkey dinner for our closest friends.

My cousin Scott married just about the sweetest person you'll ever meet—Helen Young. At that first turkey dinner, she brought a squash dish that was the hit of the night. Each year after that (until Latran ceased to exist and the wellspring of free turkeys dried up), we demanded in no uncertain terms that she bring her famous squash dish again and again. Helen was kind of enough to share the recipe. And so, with full credit given to the squash maestro, here is Helen's famous squash dish.

The nuts and brown sugar on top are the key to making a simple vegetable dish come to life. Note that you can give this dish a nice kick in the ass by adding some cooked bacon and kale (both finely chopped) to the mixture before you add the topping. Serve with a bit of maple syrup if you're in a really decadent mood.

Helen's Thanksgiving Squash

Ingredients

3 cups cooked butternut squash (about 1 whole big squash or two little ones - boil till soft then mash)
1/2 cup milk
1/2 stick butter
1/4 teaspoon ginger OR pumpkin pie spice OR cinnamon
1/4 teaspoon nutmeg
1 tablespoon vanilla
2 eggs
Optional: kale and/or cooked bacon, both finely chopped.

Steps

1. Mix all the above ingredients together while the squash is still warm. Pour into a greased casserole dish.
2. Then mix the following together to make the topping, and sprinkle on top of the squash mixture:
   1/2 cup brown sugar
   1/3 cup flour
   1/2 stick butter
   1 cup chopped walnuts
3. Bake at 350° F (177°C) for 35 minutes

Pour that thick orange liquid creation into a baking dish (left). Top it with the powdered nut confection (right).

After baking, you have a nice caramelized topping.

Recipe 12

Dan's Incredible Baked Beans

Not to take any of the well-deserved spotlight away from Helen… but let's divert some of those photons of praise onto a different recipient. This is Dan Peduzzi's fourth mention in this book. So, you should remember him by now. Well, he brought a batch of baked beans to one of our parties and I have to admit I was shocked at how damn good they were. So good they almost make your knees buckle—like the cast of Seinfeld experiencing that unbelievable "Soup Nazi" soup. (Bite… chew… chew… *oh, my God… I have to sit down.*)

They're a bit labor intensive to prepare, so I must admit I've never actually made them myself. But I've experienced their delight on several occasions. Accordingly, let's ask the bean-master himself what he has to say about the best damn baked beans you'll ever taste.

Dan says…

I was given this recipe many years ago when I was just out of college and on my own. It represents my first step beyond the 3- or 4-ingredient meals to which I had become accustomed. Over the years, it has served as a reminder that food preparation can be more than just a means to an end; that the experience leading up to consumption can also be fun, relaxing and satisfying. This recipe never fails to impress!

Dan's Incredible Baked Beans

Ingredients

16 oz dried kidney or navy beans
3-4 cinnamon sticks
1 large onion (or 2 medium)
6 cloves garlic, pressed
1 kielbasa, sliced and chopped
2 cups chopped beef or pork
3 cups liquid from simmered beans
1/2 cup corn syrup
1/2 cup dark molasses
1/2 cup white vinegar
1/2 cup ketchup
2 tablespoons dry mustard
3-4 tablespoons brown sugar
16 oz can chunk pineapple, drained
1 tablespoon salt
2 smoked ham-hocks or 1 hambone (optional)

Steps

1. Place dried beans in large pot on stove and cover with 3-4 inches cold water. Let soak overnight (8 hours or so.) Drain and rinse beans, discarding imperfect beans and/or small stones.
2. Return beans to pot, cover with 1 inch of water. Add cinnamon sticks (you can also add 8 or so whole allspice/dried juniper berries if you like.) Bring to a boil, then lower heat and simmer beans for 45 minutes to 1 hour. Drain beans, saving approximately 3 cups of liquid and all cinnamon sticks, etc.
3. Sauté onion and garlic in 1-2 tablespoons oil in saucepan for 5-10 minutes or until tender—do not brown. If you do not own a garlic press, peel cloves and squash flat with the bottom of a drinking glass, then chop. Add kielbasa, sauté for 10 minutes or so.
4. Combine corn syrup, molasses, vinegar, ketchup, dry mustard, and brown sugar in crock pot. Stir until combined evenly.
5. Add ham hocks/ham bone to crock pot. Combine beans, onion/garlic/kielbasa mixture and chopped beef or pork to crock pot. Stir, keeping ham hocks/ham bone buried in mixture. Add salt and pineapple, stir. Include cinnamon sticks. Add bean liquid as necessary to just cover entire mixture.
6. Cover and cook on low 10 to 12 hours. (High: 4 to 5 hours, stirring occasionally.)
7. Before serving, remove any cinnamon sticks that remain intact.

Recipe 13

Time for Cocktails!

A number of years ago, Julie and I discovered that a hotel bar in our town had an amazing selection of specialty cocktails. We ended up making it a weekly ritual to go there on Fridays after work and each have our one drink for the night. (I know, we're total lightweights.) Our favorite bartender, Shevaun, was so friendly and made such great drinks that we eventually asked if we could hire her to make all these drinks at our annual Xmas party. She agreed and I was given a list of all the bottles and ingredients I'd need depending on which cocktails we wanted to have available.

And now, with a simple trip to the liquor store, you too can obtain what's required to make some of the best cocktails for a party (or just for yourself if you're like George Thorogood). The most important thing to tell you is that when it comes to the chocolate liqueurs you absolutely *must* buy the Godiva brand. I'm serious about this. If you go with anything else you *will* be disappointed. Trust me on this one. Buy both the Godiva *dark* chocolate and Godiva *white* chocolate liqueurs. They're expensive, but they're worth it. For the rest of the flavored vodkas and liqueurs you can grab whichever brand is on sale. The only exception is the Elderflower liqueur. If you want to make a pear martini, you have to get the St. Germain brand. Again, expensive but worth the cost. And if you want to give the Lemon Drop an extra lemony boost, feel free to pick up a bottle of Limoncello liqueur and add a shot to the recipe below.

As far as the mixers go, here's a few tips. *Pom* brand pomegranate juice is crazy expensive. There's a brand called *Langer's,* though, that makes a Pomegranate/Blueberry in a 100% juice variety at half the price. We also really like several of the juices we get at Trader Joe's. They make a great Blackberry Crush, Pear, Blueberry and a "Dixie Peach" juice that is so good that when you take a swig of it, it's like someone mashed a fresh, ripe peach right in your mouth. With the milks, if you don't have chocolate milk on hand, you can use any milk. I like the unsweetened vanilla almond milk for most drinks.

Stocked up and ready to get shaking.

And with that said, here's a selection of some of our favorites. I've increased the ratio of the mixers to the alcohol from the original recipes. If you like a stronger drink, you can obviously just use less juice or milk when you mix them up. These quantities are good though, if you want to give your guests the chance to try a few different drinks without getting too sloshed. When making for just us, I use even more juice. And then instead of dumping all the ice cubes from the shaker, I add a few of them to the glasses so we can take our time sipping them and keep them nice and chilled.

Note that the Gingerbread flavored coffee creamer has been tough to locate these past few years. It may have been discontinued. I haven't found an equivalent substitute yet, so hopefully you'll have better luck finding it (or a good replacement). Also, make sure you don't mix up the *white* and *dark* chocolate liqueurs and the creams de cacao. It can turn a bright and cheery-colored drink (like the Peppermint one) into a dark and gloomy looking cauldron brew if you use the dark instead of the white by mistake.

And now you're ready to enjoy this selection of fun tasty concoctions. For added effect, add whipped cream (from a can) for the dessert drinks, and cherries to top some of them off. My favorites are the Banana Cream Pie, the Godivatini and the Chocolate Raspberry-tini. Julie loves the Peppermint, the Pear Elderflower and the Pumpkin Pie martinis. Sláinte Mhath!

Banana Cream Pie Martini

½ oz Light Rum
1 oz Crème de Banana
½ oz **White** Crème de Cacao
1 oz Godiva **White** Chocolate
2 ½ oz Milk

Pumpkin Pie Martini

½ oz Vanilla Vodka
1 oz Godiva **White** Chocolate
½ oz **White** Crème de Cacao
1 oz Gingerbread or Pumpkin Creamer
2 Dashes of Pumpkin Pie Spice
3 oz Milk

Appletini

1 oz Vodka
1 ½ oz Apple Pucker
3 oz Apple or Cran-Apple Juice
Add Cherry and Splash of Cherry Juice (Optional)

Peppermint Xmas Martini

½ oz Vanilla Vodka
¼ oz Peppermint Schnapps (or Just a splash)
½ oz **White** Crème de Cacao
½ oz Green Crème de Menthe
½ oz Godiva **White** Chocolate
3 oz Milk

Godivatini

1 oz Vanilla Vodka
1 oz Godiva **White** Chocolate
3 oz Milk

Chocolate Raspberry-tini

1 oz Raspberry Vodka
1 oz Godiva Dark Chocolate
½ oz Godiva White Chocolate
½ oz Raspberry Liqueur
2 ½ oz Chocolate Milk

Pomegranate Martini

1 oz Pomegranate Vodka
½ oz Pomegranate Liqueur
3 ½ oz Pomegranate Blueberry Juice

Raspberry/Blackberry Martini

1 oz Raspberry Vodka
½ oz Raspberry Liqueur
3 ½ oz Pomegranate Blueberry Juice or…
3 ½ oz Blackberry Crush Juice

Lemon Drop Martini

2 oz Citron Vodka
1 oz Cointreau or Triple Sec
1 ½ oz Fresh Lemon Juice
½ oz Simple Syrup
1 oz Limoncello (optional)
Lemon Sugar Rim

Caribbean Martini

1 oz Malibu Rum
½ oz Vanilla Vodka
½ oz Crème de Banana
3 oz Assorted Juices (Pineapple, Orange, Peach…)

Mudslide Martini

½ oz Vanilla Vodka
½ oz Kahlua
½ oz Bailey's Irish Cream
½ oz **Dark** Crème de Cacao
3 oz Milk

Egg Nog Martini

½ oz Vanilla Vodka
½ oz Spiced Rum
½ oz Amaretto
3 ½ oz Lite Egg Nog
Dash of Cinnamon/Nutmeg

S'mores Martini

1 oz Marshmallow Vodka
½ oz **Dark** Crème de Cacao
½ oz Godiva **Dark** Chocolate
2 ½ oz Milk
1 oz Gingerbread Creamer

Pear Elderflower Martini

1 ½ oz Pear Vodka
1 ½ oz Elderflower Liqueur
2 ½ oz Pear or Blueberry Juice
Add Pear Slice/Lemon Twist
Use a Lemon Wedge and Fine Sugar to coat the Rim

Recipe 14

Hanlon Sugar Cookies

Well, you can't have a book of recipes and not have anything for dessert. And who doesn't love cookies? I know I do. There's nothing like fresh baked Toll House cookies fresh out of the oven. Yet, there's so many recipes out there for chocolate chip or oatmeal raisin or peanut butter cookies. So, there's really no need to include another version of those same cookies here. Don't get me wrong. I absolutely *love* those cookies more than just about anything. But here's a cookie recipe that you might not have, that has a special meaning to us.

Each Christmas season, I am required (in no uncertain terms) to assist in the baking of the Hanlon family sugar cookies—and then in the decorating of these cookies with homemade frosting of various colors. This is a chore I pretend to dread. But there's something about the perfect buttery taste of these cookies. And the homemade frosting gets that just *slightly* hard quality on the surface, but still has that smooth, sweet, creamy texture underneath. Cookie perfection. Once again, I'll turn it over to Jules.

Julie says...

Sugar cookies... I know. Boring right? WRONG. This is my mom's recipe and it is so simple, yet the very best sugar cookie I have ever had. We make these every Christmas. The recipe card below is in my mom, Jean's, beautiful handwriting. My mom has been gone 30 years now (November 10, 1990). Some of you may remember her. She was and always will be my very best friend. My mom was the definition of strength, beauty, kindness, intelligence and humor. Growing up, my mother never had to tell us she loved us, because she showed us every day. I am so proud of many things in my life, but at the top of the list, is that I am Jean Hanlon's daughter.

My Mom loved Jeff. Yep... they knew each other. She loved how smart he was. And that he was so confident (or maybe a bit arrogant...) that he would do the Boston Globe crossword every day in pen. On Sundays when Jeff would call me, Mom would say, "Ask him how far along he is on the crossword." So I would. Jeff would say something like, "Pretty good. I'm about 85% done." My Mom would quickly say, "Tell him I'm 90% done." Each and every time she would be ahead of him. Or at least that is what I was told to say. In truth... she was always behind him, but she liked to keep him in check. He never knew that until after she died. Well played, Mom. Well played.

Two points of clarification: Firstly... So yes, I did do the crossword in pen. But that wasn't really an arrogance thing (I fully admit I was an arrogant little shit back then, in many ways). It was because I hated the way the pencil lead would smudge as you drag your hand across the crossword grid over and over again. Ink dries quickly and stays put. And if you don't make any mistakes, you don't miss not having an *erase* option. (Hmmm… maybe it *was* a little arrogant of me then. Ha.)

And secondly… You would think after years of Jean Hanlon always being several percentage points ahead of me in the Sunday crossword week after week after week, I would've figured out that she was just constantly lying about it. Nope… Guess I wasn't that smart after all, was I? I tip my hat to you, Mrs. H, you *liar* you.

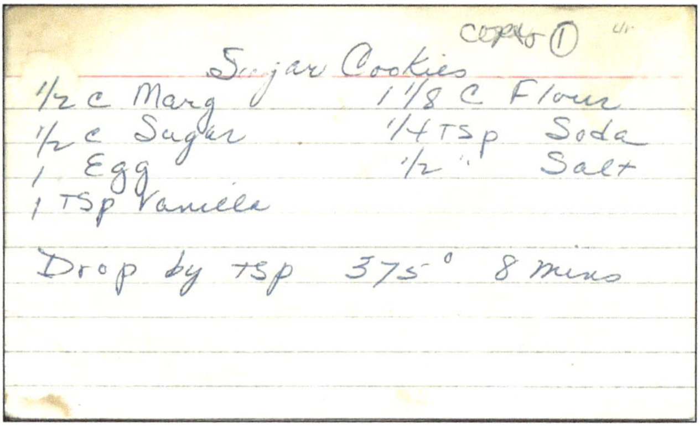

Now let's type out that recipe in case you're not sure of the abbreviations. And of course, we use real butter now, not margarine:

Ingredients

½ Cup Butter, softened
½ Cup Sugar
1 Egg
1 Teaspoon Vanilla Extract
1⅛ Cup Flour
¼ Teaspoon Baking Soda
½ Teaspoon Salt

Steps

1. Mix everything together at once in a large bowl, either by hand or with a mixer.
2. Bake at 375°F (191°C) for 8 minutes.

We don't spoon out the batter onto the baking sheet to make the cookies, like it says on the recipe card. We spread some flour on the counter (and also on the rolling pin) and flatten out the ball of dough. Then we use our cookie cutters in various holiday shapes to make holiday cookies in the forms of snowflakes, xmas trees, sleighs, candy canes, holly, etc. …

We usually roll out the dough to be about ¼" thick. If you go down to only ⅛" thickness, the cookie will be all *crisp* and no *chew*. You may like that. We like it when they're chewier. (What happens if you go up to ⅜" or ½"? I don't know. Are you mad? *You'll kill us all!*)

Now collect up all the excess dough and roll it out again (with more flour to prevent sticking if necessary). Keep repeating the process until you've used it all up. I usually take the last of the scraps and make one un-shaped "blob" cookie for immediate consumption straight out of the oven.

You want the dough cold when you do all of this. So put the dough back in the fridge for a bit if it gets too sticky. Note that if you make a double batch, it's good to rotate the doughs in and out of the fridge as you cut each batch of cookie shapes and bake. Julie usually makes the dough the day before and keeps it in the fridge. Sometimes she makes it up to a week in advance and keeps it in the freezer until the day before, and then thaws it in the fridge.

You can cook these directly on a cookie sheet. If you use parchment paper, you'll improve the evenness of the cooking, and reduce the chances of burning the bottoms of the cookies. Or alternately, for a very reasonable price you can buy some silicone non-stick "baking mats." These are great. I highly recommend. Work as well as parchment paper. Easy to wash. Totally reusable. So much cheaper than parchment (which is pricy) in the long run.

I bought a set of two Gridmann mats on Amazon for only $11. Just make sure you bake them once by themselves before you use them. They'll stink a little as they outgas a bit during that initial cycle to high temp and back. After that, they're good to go and can be used for many more baking jobs than just cookies. You can roast veggies on them or bake chicken and easily wash the mat when you're done.

I know. Those aren't Christmas cookies (left). Sorry. We made these in the springtime and therefore, we weren't allowed to make Christmas shapes—even if I wanted to take a picture for the book. There are rules that simply can't be broken. Note how the bottom right edges of those butterflies look a little crispy. Those sections were a little thinner than the rest. But how 'bout the great job Julie did frosting this batch (right)?... Seriously. These things are works of art.

Recipe 15

Mark Hanlon Memorial Frosting

The frosting itself is pretty basic. Just four ingredients (if you don't count the food colorings). Once you've made a batch of frosting, you can separate it into four or five different containers and use food coloring to achieve several shades of decorative frosting. We usually have red and green for Christmas-shaped cookies. We use a small amount of blue coloring to get a sky blue that looks great (combined with uncolored white) on the snowflake-shaped cookies. And we have a few shakers of colored sugar crystals that work well when used sparingly. Again, here's Jules with the story behind the music…

Julie says…

The "Mark Hanlon Famous Frosting." Mark was my brother who passed away in 2006. He was an incredible big brother. He was 11 years older and, in a way, was my rock. As some of you know, my Dad got sick when I was only 2. My older brothers were saddled with taking care of me a lot when my Mom would be at the hospital. Mark took me everywhere with him. Every Saturday morning, he would take me on errands. We would get donuts, and go to Salem, New Hampshire, to the Pizza Hut for lunch. He made me feel like I was the coolest kid around. (Surprise… I wasn't!) I have so many great memories of being in his orange Honda, cranking the radio and driving around. He never made me feel like a bother. I digress… he was also an awesome cook/baker.

This frosting is famous amongst the Hanlon crowd. We have it on birthday cakes, cookies, cupcakes… And for years (when I would ask Mark for the recipe), he told me it was a secret. It wasn't until after he died (when my nieces Maggie and Molly took over the birthday cake-making duties from their Dad), that they told me the "famous" frosting recipe was from the Domino sugar box.

For years, I thought this was a magical frosting. And yet, it came from the box. But as with most things Mark… it holds such wonderful memories. The frosting (and everything he did) was magical in his little sister's mind. In addition to being Jean Hanlon's daughter, I am Mark Hanlon's favorite (and only) sister. I'm very proud of that also.

Ingredients

4 cups confectioners' sugar
1 cup unsalted butter, softened
4 tablespoons whole milk
1½ tablespoons pure vanilla extract
Food coloring, various colors (if desired)

Steps

1. Mix the sugar, butter and vanilla together; continue mixing while adding milk slowly until the frosting is as smooth as desired. Julie uses a mixer on low setting and then increases it to medium. Add food coloring to tint the frosting to the color desired.

Luckily, my niece, the Marvelous Mrs. Maggie (Hanlon) Surette, made a batch of actual Xmas cookies this year and sent me a pic (above, left). Now that's the good stuff. Thank you, Maggie. Nicely done. But since when is yellow a Christmas color?! Ha.

My other niece, Molly (Hanlon) Burr, also came through for me (above, right), but a certain 2-year-old (below, left) must have eaten most of them before I could ask for a pic. I'm on to you, Gavin! And there's my best friend, Grace (below, right), having the time of her life making a delicious mess of the whole process.

Recipe 16

Recipe for a Better Body

I hope at least one of those recipes ended up pleasing your palate. If not, my humblest apologies. The sense of taste is subjective. There's a chance we may not be on the same wavelength. The visible spectrum goes from red to violet. Has anyone ever created a taste spectrum? Call me part indigo, part red. And what the hell is indigo? I mean, just look at the visible spectrum. There's blue. There's violet. You're saying there's a whole 'nother color in between those two? I think we're just talking shades of purple, people. Besides, without indigo the ROYGBIV spectrum is reduced to the perfect symmetry of just the three primary (RYB) and secondary (OGV) colors. [R+Y=O, Y+B=G & R+B=V]. So take your tertiary butt and step to the side, Indigo! Wait, what happened? Did I get sidetracked again? Sorry. OK. Back on topic...

Around the year 2001, my aging body, my slowing metabolism and my sedentary lifestyle finally caught up with me. My weight had slowly risen up into the high 170's—a pretty depressing statistic for a guy who measures under five and a half feet. When that happened, I finally made a commitment to get in shape. Over the course of the next year and a half, I denied myself the pleasures of eating the foods I liked and I jogged about a mile every other day on average. I lost 38 pounds in about 18 months. That was without a doubt the hardest thing I've ever done. Ever.

I briefly hit the 130's (138.8 to be exact) and said "*Mission Accomplished.*" I kept the weight off for a number of years (hovering in the 140's with the normal fluctuations). Then I put it all back on in the late 2000's. I was working at a start-up company which demanded I work *way* more hours than they were paying me for. (We won't get into the unfulfilled promises and the bad decisions that led to that scenario.) Because of that, I was stress eating and I wasn't getting any of the regular exercise I had added to my life previously (mostly long walks) due to my exhaustion and a lack of available time. My weight once again shot up into the 170's. *Depressing...*

I couldn't believe I was going to have to do it all over again. Yet with a dejected acceptance, I agreed with myself to suck it up and start slowly taking off the pounds once more. I had successfully dropped over a dozen pounds the hard way when I came across a book. I had borrowed the Kindle from my local library for a trip to Turks & Caicos. I used to throw 3 or 4 books in my suitcase every time we went on a week's vacation. But this time, I had the library Kindle which was pre-loaded with over 100 new releases.

Take advantage of your local library! Libraries are great. I used to display all the books I had read like trophies in my bookcases. Finally, I started using the library and then never looked back. Now I get 98% of my books from them. We also get video rentals, audiobooks, and e-books for free. On top of all that, our library has equipment you can borrow, like the aforementioned Kindle. I've saved thousands of dollars over the past 20 years in book costs and DVD rentals. Again—libraries are great.

So, as I lay on the beach under the shade of a well-placed umbrella, I read. (Yes. Under the umbrella. I'm there for the swimming and the warmth, *not* the skin cancer.) After finishing one book on that Kindle, I scanned what else was loaded, between dips in the gorgeous cyan-blue waters of the Caribbean. I came across "The 4-Hour Body" by Tim Ferriss.

Turns out he has a great regimen for weight loss that ended up working really well for me. I highly recommend the book. In addition to the section on weight loss, he has chapters devoted to building muscle mass, better sleep, better sex... It's full of great info. Good thing I had the e-book, 'cause the actual physical book is heavy as shit, I later learned.

The thing that I liked about it is how he treated his body like a science experiment. As an engineer, I got a kick out of this. He has a weight loss method called the Slow Carb diet. It's nothing revolutionary. It just worked for me. And it was the first time I saw a diet that didn't require you to limit how much you eat. It really is all about *what* you eat and *when* you eat it.

He also had some nice shortcuts. For example, drinking water is great. We all know that. But we mostly forget that a calorie is a measure of heat. Therefore, if you drink a lot of *ice-cold* water, your body has to burn energy (calories) to heat that water up. He then proceeded to take it a step further and measure his body's response to cold showers and then in extreme measures—*ice baths*. Yikes. Like I said, he's a one-man science experiment where he was his own test subject.

Here's another useful trick. Using Pareto's law of getting 80% of the benefit from 20% of the effort, he details how you can cut down on your exercise by timing it properly. If you end up doing as little as two minutes of exercise before you eat a meal, you can open up the receptors in your muscles to devote more of the fuel you take in to building muscle instead of fat. The more muscle you have, the more calories that muscle mass burns without any extra effort on your part. Put several of these little things together, and they add up to incrementally easier weight loss.

Overall, I only did a half-assed version of what he recommended. I *reduced* the amount of forbidden foods I ate, instead of eliminating them completely. I mostly followed the protocol though. I ate breakfast every morning that was high in protein (eggs usually or smoothies with vanilla protein powder). This was a change for me, since I had usually skipped breakfast, not realizing that my body was becoming hypoglycemic when I did that—later reacting to that by going into fat storage mode when I finally gave it some fuel.

I ate a bigger dinner, and I did tiny exercise routines before I ate. (Following Tim's lead, I would even do air squats in the men's room of a restaurant when eating out. Laugh all you want. It worked for me!) I used to prepare a regular-sized dinner. Then shortly afterwards I'd have the urge to snack. That's because I didn't make a big enough meal. Put more veggies on that plate—*more than you think you want*. You'll end up eating them all. Then you'll be less likely to snack afterwards. The best veggies are spinach, broccoli, cauliflower, kale, asparagus, green beans and black, red & pinto beans. (Just read the "Slow Carb" section of the book).

I also chugged lots of ice water. And in the summer months I would reduce the temp on my showers in steps. Each time I adjusted the water a little colder I would get that cold shock and think, *it's too cold!* Then I'd adjust to it within a few seconds like when you jump in a cold pool. Then I'd drop the temp again. *Too cold!* Equilibrate… Temp drop. *Too cold!* Equilibrate… Temp drop. *Too cold!* Often, I would get to the point where it was pure cold coming out of the shower head and I'd shriek like a little girl with that last initial freeze shock. Then I'd get used to that as well, and let the water suck the heat out of my body by conduction for several refreshingly cool minutes.

Also, look up how to do a quick body fat estimate. If you lose only 20 pounds, but you also convert 10 pounds of fat into muscle mass, then you've actually lost 30 pounds of body fat. I wish I had done that measurement before I lost my weight. I'd love to know what the total amount of body fat was that I actually lost.

Long story slightly shorter, I lost a total of **36 pounds** and hit an adult-low for me of **136.6 lbs.** I joked that I did this by *eating more and exercising less*. This was a true statement since I had added a regular breakfast to my daily food intake, and exercised way less than the first time I lost the weight.

WeightWatchers has a pretty good system too with their "zero-point" foods to avoid starving yourself. But this one worked for me. So that's why I recommend it. Note that the "Cheat Day" once a week works *only* if you follow the protocol. You spike your metabolism once a week by eating anything you want. That keeps your metabolism from slowing down too much and working against you as you try to lose more weight. If you don't follow the protocol though, the cheat day will screw you over *big time*. So that day after "Cheat Day" is the most important day to stick to the recommended foods and not half-ass it.

"A word is only worth a milli-picture." (A brilliantly clever phrase that my friend Jim Zambuto originally coined.) So let's take a look at a visual representation of the weight loss pattern I'm talking about (conveying those proverbial 1,000 words).

This chart shows my weight tracking from that second time I gained the weight through to that new "adult low." I started the 4-Hour Body regimen after I had already lost the first 15 pounds or so the hard way. At that time, I started charting my daily

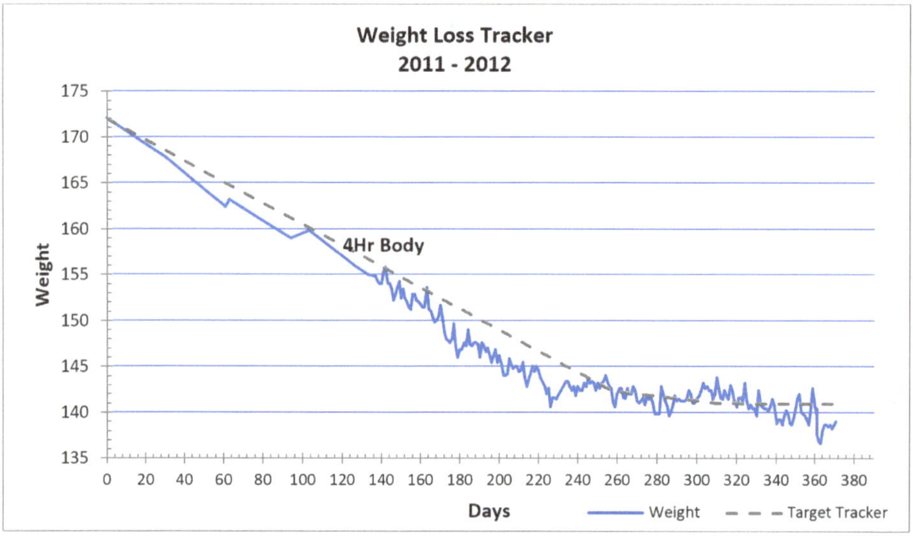

weight. From that point forward, it's interesting to see how consistently my weight would pop up every week (the day after the Cheat Day). Yet a day or two later, it would fall off and then additional pounds would follow.

I definitely recommend starting a spreadsheet and tracking your weight every day. Measure yourself every morning and input that number into your chart. It's been shown over and over that the simple act of tracking your weight in this manner will program your subconscious to make better choices and help you achieve the goal of losing pounds. It's also interesting to see how your body responds day to day to what you've eaten. One example is how high-sodium food will spike my chart a pound higher due to the resulting water retention. Then a day or two later it comes off.

You can just input the numbers and do no analysis whatsoever. Or like me, you can watch and learn from the weight gain and loss—drawing conclusions from correlations that consistently manifest themselves. I love data. But again, just the act of charting your weight each day will help, even if you do nothing with it.

If you're not proficient at creating charts and plots, do a search for Four Hour Body Weight Loss Chart and you'll find a good one on his site you can use. He calls it the Glide Path.

Thanks, Tim. Your efforts are much appreciated.

And the sun on the equator, setting like an ember thrown to deep water
From crimson to black, but coming back, tomorrow... on the horizon.

- Lyrics by Steve Hogarth

Dedications

Dedication #1

This book is dedicated individually to every single person who contributed to the Marillion Tour Fund in 1997. Each and every one of you has my gratitude, appreciation and friendship. We only succeeded in making it happen because we all made it happen together. This book wouldn't exist without all of you.

Insert List of 1,200 names.

(Author's Note: I have not been able to find a complete list of all the donors. You all know who you are, though. And I thank you.)

Dedication #2

On a personal note, I'd also like to dedicate this book to every music fan I've met over the course of the past 30 to 40 years. Some of you became lifelong friends. Some of you were only fleeting acquaintances. However, all of you brought a certain amount of joy into my life and I thank you for that.

This is by no means a complete list. There's a lot of you out there for which I'm only remembering first names or nicknames, and some I've forgotten completely. I hate that that's the case. Drop me a line if our paths crossed and I left you off this list. If there's ever a second printing of this book, I'll be sure to add you in. With that said, here's to you, guys…

Jim Alger
Rick Armstrong
Romy Bahri
Lee Bassage
Denise Bouffard
Lori Burr
Jim Clark
Jon Collins
Rick Collins
Rob Crossland
Mark Dadgar
Andreas Dahl
Marina Dannis
Jean-Francois Jeff Denis
Seth Dewey
Rita Drew
David Dyte
Kristie English
Scott Forte
Erik Foxvog
Rob Galardi
Joann Garrido
Jeff Gebhart
Matt Gehrke

Pete Gilbert
Myron Grubaugh
Malcolm Harris
Cynthia Hetherington
Lasse Hoile
Christine Holz
Stephen Humphries & Kim Proctor
Geoff Kieffer
David Levine
Cathy Loomis & Brian Steed
Annie Lutz
Shane McKie
Lisa Mikita
Rys Miller
Bill Novak
Lisa Ondrey
Derek & Jeanine Perduta
Polly Peterson
Carmen Quintero
Marlies Raadersma
JR Remke
Kim Robinson
Jonathan Rose
Deb, Cara & Andy Saks
Kim Salter
Jeff Schwartz
Len Senecal
Marko Simic
Phil Simon
Stephanie Stago
Frank Sullivan
Jodi Leigh Tack
Gary Uzzell
Tom Vance
Andy Verdi
Diana Brie-Wharton
Cynthia Wright
Carol from PA
Chris & Kim (military couple stationed in the UK)
Max & Marie Luisa from Italy

Mr. Andy Saks and I partaking in some silliness at the Ben & Jerry's ice cream factory in Vermont (top left), on our way home from a Marillion Weekend convention in Montreal. The Minehead convention gang (top right) in 2005. Hanging backstage with the lovely Kim R (bottom left) and new friend Dave L (bottom right) in Montreal.

Dedication #3

This book is of course dedicated to each of the members of the band Marillion. I honestly don't know how they can all be so talented, and so celebrated, and yet still be some of the nicest guys you'll ever meet. Over and over again, through the many years we've all known them, they consistently make themselves available to their fans. They give of their time and of themselves. And let's face it, some of us truly are freaks. How is it even possible that they hold conventions with a couple thousand of their most *obsessed* followers and they don't even hire any security?

Thanks for the inspiration, thanks for your generosity and most of all, thanks for the music.

Steve Hogarth—Vocals, piano, and impish grins.
Mark Kelly—Keyboards, samples, and maniacal plans
Ian Mosely—Drums, percussion, and presidential wisdom
Steve Rothery—Guitars and godlike, instrumental powers
Pete Trewavas—Bass, guitars, and onstage bouncing with boyish, joyful smiles

Taken by me after being invited to watch the soundcheck (left) in Montreal at the 2019 Marillion Weekend. Surprisingly, the soundcheck wasn't just pieces of songs. They played full songs for over an hour with the same lights as the show that night. It was like our own personal concert with no one else in attendance. Thanks, guys! And confetti explosions for the finale (right) as seen from the balcony.

Dedication #5

Many thanks also go to Marillion's supporting cast. You guys have certainly shown Julie and me many kindnesses over the years. Thank you. It hasn't gone unnoticed, nor unappreciated.

Stephanie Bradley, Lucy Jordache, Mark Kennedy, Richard Lee, Erik Nielsen, Ant Short, John Wesley and many others.

Dedication #8

I'd like to additionally thank some other music people who have given me and the world some great joy. In no particular order…

Peter Gabriel, Genesis, Derek "Fish" Dick, Joe Jackson, Kevin Gilbert, Steven Wilson, Porcupine Tree (*How did I write an entire book about music and not mention SW and PT? Thanks for giving my passion for music a much-needed figurative shot of adrenaline about 20 years ago right when it needed it most!*), Annie Lennox, Roger Waters, Pink Floyd, Paul Simon, Art Garfunkel, Cat Yusuf Stevens, Alan Parsons, October Project (w/Mary Fahl only), Moya Brennan, Clannad, U2, The Who, The Doors, The Cars, Neil Finn, Jeff Lynne, Glen Hansard, Donald Fagen, Walter Becker, R.E.M., Rush, Kansas, Bruce Springsteen, Supertramp, Sting, The Police, Yes and (of course) The Beatles.

The Amazing Amy Sullivan and me, hanging with our posse – aka R.E.M. Finally got to meet them in Boston (after not being able to in Milan). Yes. That's the incredible Mike Mills gripping my shoulders – perhaps in solidarity for our similar outfits? Ha.

Dedication #13

This book is dedicated in large part to my dad, who's been gone for 10 years now. I wish that I had matured fast enough that our relationship could have been so much closer. How unjust that I directed so much of my frustration with the unfairness of the world towards the guy that stuck around through it all.

My dad was a mechanical engineer, a licensed & instrument rated pilot & Piper Comanche plane owner, a graduate level teacher at Union College, a licensed real estate broker, an Eagle Scout & later a Scout Troop leader, an owner & manager of multiple businesses, an apartment building property owner, an incredibly talented amateur musician on guitar & piano, a great singer, a Little League baseball manager, a world traveler, a member of Mensa, a ballroom dance instructor at Arthur Murray Studios, a formidable opponent on the tennis court and basically an all-around well-respected and incredibly-liked guy. And also, in hindsight, a pretty damn great dad. If there is in fact another plane of existence, I hope this message of love and appreciation echoes through. And know this... Your waffle recipe will live on, Dad.

Dedication #21

This is also dedicated to what remains of my childhood family. I thank all of you for your role in helping me become the man I am today. Ellie & Rick Chappell (*thanks for the great recipes, Mom. And, of course, for giving me life.*), Ceil & Peter Eisenhower (*Thanks, Ceil, for giving me my first album at the age of 10 [Fleetwood Mac's* ***Rumours****] and starting me on my journey of musical discovery & appreciation that dominated most of my life and resulted in much of what was documented in this book today. You're a great lady. And you always were the "Cool Aunt."*), Alan & Venessa Pelletier, Dan Pelletier (*who actually bought the* ***Afraid of Sunlight*** *CD even if it was only to appreciate {as a Francophile} the French spoken word intro to "Beautiful"*), and Elizabeth Pelletier.

Dedication #34

This also goes out to my "new" family. Molly & Ryan Burr (and Gavin & Addison), Lynne Cox (*Thanks for the scholarship, Lynne! I hope I proved worthy.*), Amanda & Pat Hanlon (and Mateo) (*You made the final chapters of this book happen, Amanda. Thank you again!*), Chris Hanlon (*keep discovering Genesis and more of those great classic rock bands, Chris!*), Cindy Hanlon, Mike & Marcia Hanlon (*Great job on the Tribune article, Mike!*), Steve Hanlon, Tim Hanlon, Chris & Maggie Surette (and Colin & Kate).

It is a unique and special thing to see a family that loves each other as much as you guys do. Keep up the great work.

Dedication #55

An extra-special thanks goes out to Mike Hanlon from the previous list. This book exists in its printable form because of the many hours that Mike spent editing and converting my draft into Adobe *InDesign* to be "printer-ready." Thanks, Mike. I don't think I'll be ever be able to repay you for your generosity, but I promise to try. Thanks also to Dan Peduzzi, Jon Collins, Stephen Humphries and especially Phil Simon for their first looks and feedback on the initial drafts of this book. The input and corrections each of you provided absolutely made this a better book. Thanks also for the praise and for the honest criticism (in the appropriate ratio). The blame for any remaining errors rests solely with me.

Jeez, Jeff. How many dedications are you going to have? You're only supposed to have one. Aren't you? The answer to that, of course, is "42"—the answer to all questions, including the eternal question of life, the universe and everything.

No, really. Who says you have to do things the way they're always done? Break the rules if it doesn't hurt anyone. And so, I'm going to thank everyone I can think of. Because I think it's kind of cool to be thanked by name in a book. And if doing so brings a tiny bit of happiness to these people (even the ones that are too cool for school to admit it), I'm not letting the opportunity to do that pass by. And accordingly, the Thank You Parade continues!...

Dedication #89

Thanks to all the friends that have been a part of my life from as early as grade school. Your friendship has been a constant in my life that I will always appreciate. I've learned that most people don't have friends from their high school years. They all re-format their friend hard drives and start over with a clean slate in college. I'm so glad that didn't turn out to be the case with us. Many of you are quite simply the most important people in my life. I'm so happy you are still with me on this journey. I love you, fully and unabashedly.

Melissa (Witty) Zampitella, Scott & Helen Young, Hilary (Carrick) Waldron, Jimmy & Joyce (Deroch) Thibadou, Rich Smith, Lauren (Hastings) Smith, Justyn

(Scarbeau) Smith, Lorraine (Coughlin) Sabella, Robert Reed, Karyn (Worthley) Pierog, Mark Perlberg, Kevin Levesque, Steve Boush, Mike Boush, Donna Bourgeois.

You guys are the best! Every one of you.

Dedication #144

While we're at it, there were many more great people from those early years. They may not be part of my life now, but I have nothing but fond memories of all of them. So, let's remember them too. This book is also dedicated to the distant or not-so-distant memory of the following people, just to name a few.

Joellen (Spencer) Adams, Kim Beaton, Dennis Beggan, Pat Burke, Kate Carpenter, Sue Cunningham, Roger Fisk, Lisa Frazetti, Debbie Harmoning, Scott Higgins, Steve Higgins, Kurt Holmy (*the smartest guy in our high school*), Jean Kelly, Diane Klockow, Glenn Livermore, Tim Maxfield (*my best friend from elementary school*), Lisa Menard, Joe Milora, Chris Mitchell, Kathy Murphy, Eric Neiman, Paul Oshan, Sharon Parker, Susan Parker (*a great friend who named herself my surrogate mom for a while there*), Liz Parthum, Stephanie Patten, Sarah Poor (*who had the coolest taste in music long before my tastes matured beyond the standard popular rock of the day*), Jessica Powers (*she of the infamous Brechin Terrace Big Wheel race of 1970*), Sara Michelle Sapuppo (*my best friend from French class*), Robin Savoy, Amy Sirois, Amy Skinder, Kim Soucy, Kim Sullivan (*my first girlfriend—even though we never even went on a date*), Andy Thompson, Liz Walsh, Mary Walsh, Nancy Wolk, Dave Workman.

Dedication #233

Thanks to all the teachers I've had throughout my life. It really is a noble profession. And when it's done right, it can inspire greatness, or at least the desire to be great.

And that's pretty damn great, isn't it? In particular, thanks to Madame Zappala, Mr. Lynch, Mr. Lawson, Mr. King from A.H.S. Thanks to Mrs. Harrison from A.J.H.S. (*Are you still out there, Mrs. H? Do you still have a copy of that Batman play I wrote in your 9th grade English class? I wish I kept a copy of that.*) And from U. Lowell, extra special thanks to Dr. Mike Fiddy (*my graduate thesis advisor*) and Prof. Bodo Reinish (*who gave me my first job in my field at the U. Lowell Research Foundation*). I owe a lot of my success to both of you.

At grad school commencement with Prof. Reinish (left) and undergrad with fellow engineers Brian Pearson, Ed Murphy and Jeff Parsons (above).

Dedication #377

Thanks to all the friends that came into my life in my college years and continued our friendship beyond that time, most of you to the present day. My life is better because of each and every one of you. Thanks for so many great memories and thanks for the friendship. Again, those of you that are still in my life, I love you guys.

Jen Benoit, Marty & Danielle Bodley, Lisa Bodo, Wendy Bruno, Mike Burns, Janet Clark, Mike Cuomo, Greg Falvo, Alyssa (Winslow) Faulkner, Kim Gaucher, Mark Gillette, Katie Kelley, Dan Kiesling, Lorryn (Burns) Kincaid, Laureen Lens, Rick Mott, Ed Murphy, Jeff Parsons, Jill Pattavina, Brian Pearson, Dan Peduzzi (That makes *five* mentions now, Dan!), Michelle Poulin, Dan Riemer, Michelle Surprenant, Kevin Walsh, Bob Weston, Mark Winter.

Feeling the love with Dan Riemer (left) & the Smith 401 boys (right) - Dan Kiesling and Mark Gillette.

Dedication #610

Thanks to all the friends I've made in my working years. I've had some of the best times of my life with you guys by my side. Thanks for the laughs and the commiserations, the successes and the abject failures. It was a great ride.

John Zizza, Jim Zambuto, Amy (Cail) Sullivan, Linda Sniegoski, Dana Schuh, Manuela Schober, Craig Schiller, Paul Rea, Bill Nowick, Brian Nash, Marcia Murphy, Dave Morrison, Dave McCarthy, Keturah Martin, Kimny Lam, Mike Kinnear, Pat King, Kelly (Hogan) Janko, Keith Joncas, Charlie Jack, Kevin Hooper, Tom Hartai, John Hardin, Dave Goodwin, Don Foster, Pam (Gianetti) Dwinell, Glenn Coppola, Bill Clauson, Jen Clark, Tim Calnan, Alain Bouchard, Jonathan Agger.

First delivery of high-powered RGB fiber laser system for 3D movie projection at LLE.

Dedication #987

Thanks to Grace Bleckmann—my new favorite playmate and companion. You have no idea now how adorable it is that a four-year-old girl considered a fifty-year-old man to be her best friend. I hope when you're old enough to appreciate that, you'll remember our friendship and the fun we had together. Keep learning and keep laughing. "Play again next time?... *Deal!*"

Dedication #1597

Thanks to Leonardo Bonacci of Pisa (aka Fibonacci) for the numbering sequence of these Dedications—starting with the second digit so the copy editors wouldn't assume two #1's was a mistake and "correct" it.

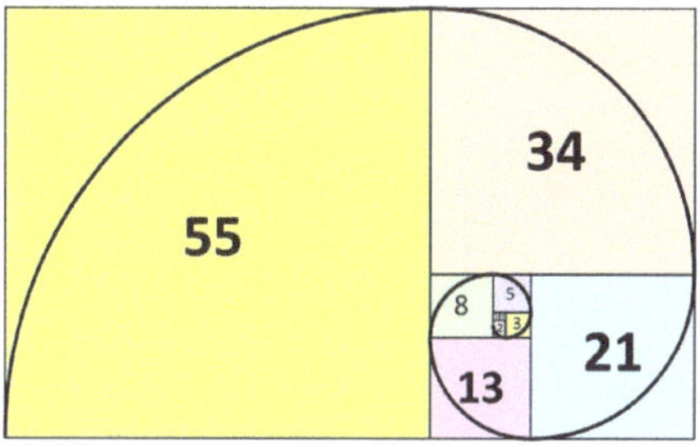

Dedication #2584

What? You thought I was going to forget Julie? I suppose that would have been OK, as I thanked her about a dozen times in these pages already and even sang her praises for half a chapter towards the end there. It's probably gotten to the point where you're sick of hearing me talk about how great she is. Well, too bad.

Thank you to literally the best person I know. You make me a better man, every day, just by being who you are. *Changing my life with a wave of your hand* (as a much, much better writer than I am once wrote–and sang.) Don't ever stop being your wonderful, happy self. You are loved as much as a person can possibly be. And yet, defying logic, not as much as you will be tomorrow.

You landed in my life, like a new and brighter light. I carry you here. And there's something of you in everything that I love.

So take my hand, Squeeze it tight. Make some light, In the darkness
I'm glad you came on this trip. Don't lose your grip, don't lose your grip
This train is our life. This train is our life. Travel with me, and we'll see...*

*- Lyrics by Steve Hogarth —*with one preposition change (from "my" to "our").*

About the Author

The author graduated from Andover West Elementary School in 1979. He worked for 30 years in the field of laser optics and currently lives in semi-retirement in a semi-quaint New England town.

Accomplishments? *Really?* Do we have to do that? Well... All right. If we have to. The author was the winner of six separate merit scholarship awards, making him one of the few fortunate people to make a profit simply by going to college his first year. The author took extra graduate courses as an undergraduate to obtain his Master's degree in only one extra year instead of two. He then spent over a year unemployed in the recession at the start of the 1990's. (The author should have taken the standard two years and enjoyed himself more in college.)

The author was a member of the *final* graduating class of the University of Lowell (before it became U. Mass. Lowell) in the *first* year they offered a master's concentration in Electro-Optics—a concentration his thesis advisor (Dr. Mike Fiddy) first created for him. Thus making him the only person on the planet with an Electro-Optics MSEE degree from U. Lowell.

The author graduated with *cum laude* honors and is a member of *Tau Beta Pi* (the National Engineering Honor Society) and *Eta Kappa Nu* (the National Electrical Engineering Honor Society). The author then worked at the defunct (and later resurrected) Polaroid Corporation for over a decade and then several start-up companies in the field of laser optics. The author has worked as a veterinarian's assistant, a gas station attendant, a water delivery person, a night manager, a research assistant, a concert usher, a covert food taster for a major burger chain, a trade show demonstrator, a laser engineer, an election worker and a writer.

The author has been to nineteen countries and been in six major bodies of water (the Atlantic, the Pacific, the Gulf of Mexico, the Caribbean Sea, the Baltic Sea and the North Sea). Note that the author only stuck a foot in the ice-cold North Sea while attending a Marillion Convention in Holland, but he still counts it.

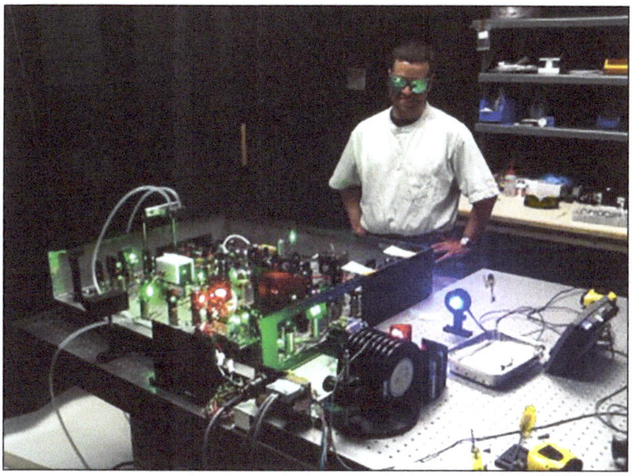

High power infrared laser producing red, green and blue wavelengths ready for fiber coupling.

The author is widely considered to be the inventor of modern internet-based crowdfunding for conceiving of the Marillion Tour Fund in 1997. The author is the writer of the book you are currently reading as well as CD and concert reviews in several paid print publications, including *The Web UK*, *Made Again* and *Music News Network*.

If you've met him, the author hopes that you will consider him one of the nicest people you've met.

The author loves ice cream, fresh-baked cookies, popcorn, breakfast, brunch, pizza, tacos, sushi, Italian cuisine, fruity cocktails, the Red Sox, Pedro Martinez, Nomar Garciaparra, the Patriots, Tom Brady, music, books, movies, great TV, a well-turned phrase, a sound night's sleep and the people in his life.

www.ingramcontent.com/pod-product-compliance
Ingram Content Group UK Ltd.
Pitfield, Milton Keynes, MK11 3LW, UK
UKHW052006201224
3805UKWH00009B/23/J
9798218161019